MORE PRAISE FOR **TAR SANDS**

....

"Andrew Nikiforuk is one of the most astute, relentless,
and original writers of his generation. This book, which reveals
the true cost of the Alberta tar sands, shows why he's so
admired in some quarters and so feared in others."
GARY STEPHEN ROSS, EDITOR-IN-CHIEF, *VANCOUVER MAGAZINE*

"This is a well-written and engaging book on one of the
most important issues of our time. If you care about global warming
and a clean energy future for our children, read *Tar Sands*."
TZEPORAH BERMAN, FOUNDER OF FORESTETHICS AND
CANADIANS FOR CLIMATE LEADERSHIP

"Andrew Nikiforuk's passionate and sobering book makes it clear
that the Alberta tar sands are more than an environmental
Armageddon—they're an essentially human tragedy. And whereas
most tragedies end with a descent into hell, this one *starts* that way."
WAYNE GRADY, AUTHOR OF *THE GREAT LAKES*

"Andrew Nikiforuk has written a simply brilliant book laying bare Canada's tectonic shift to 'petrostate.' No other work so compellingly links the ecological carnage of the tar sands to the increasing continentalism of the Security and Prosperity Partnership of North America. For any Canadian who cares about the future of our country, this is a must-read."

ELIZABETH MAY, LEADER, GREEN PARTY OF CANADA

"We have heard a lot about the oil sands lately. High greenhouse gas emissions, high energy consumption, suspected health problems downstream, a poor record on reclamation—each of these is truly troublesome in its own right. But the total picture, which Andrew Nikiforuk has assembled, reveals that we have the Guinness World Record for environmental disaster on our hands."

DAVID SCHINDLER, KILLAM MEMORIAL CHAIR AND
PROFESSOR OF ECOLOGY, UNIVERSITY OF ALBERTA

"With captivating writing and grim wit, Andrew Nikiforuk shows how the rapacious exploitation of Canada's tar sands has distorted our economy, corrupted our politics, ruined our environment, and turned us, collectively, into a rogue nation of carbon polluters."

THOMAS HOMER-DIXON, AUTHOR OF *THE UPSIDE OF DOWN*
AND CIGI CHAIR OF GLOBAL SYSTEMS, BALSILLIE SCHOOL
OF INTERNATIONAL AFFAIRS

"Andrew Nikiforuk reveals the true costs of America's oil addiction. *Tar Sands* tells an important story with passion and wit."

ELIZABETH KOLBERT, AUTHOR OF *FIELD NOTES FROM A
CATASTROPHE: MAN, NATURE, AND CLIMATE CHANGE*

TAR SANDS

DIRTY OIL *and the* FUTURE *of a* CONTINENT

ANDREW NIKIFORUK

 David Suzuki Foundation

 GREYSTONE BOOKS

D&M PUBLISHERS INC.
Vancouver/Toronto/Berkeley

Greystone Books
A division of D&M Publishers Inc.
2323 Quebec Street, Suite 201
Vancouver BC Canada V5T 4S7
www.greystonebooks.com

David Suzuki Foundation
219–2211 West 4th Avenue
Vancouver BC Canada V6K 4S2
www.davidsuzuki.org

Library and Archives Canada Cataloguing in Publication
Nikiforuk, Andrew, 1955–
Tar sands : dirty oil and the future of a continent / Andrew Nikiforuk.
Co-published by the David Suzuki Foundation.
Includes index.

ISBN 978-1-55365-407-0

1. Oil sands industry—Environmental aspects—Canada.
2. Oil sands industry—Economic aspects—Canada. 3. Oil sands—Environmental aspects—
Alberta—Fort McMurray Region. 4. Fort McMurray (Alta.)—Social
conditions. I. David Suzuki Foundation II. Title.
TD195.04N53 2008   333.8'2320971   C2008-905387-7

Editing by Barbara Pulling
Cover design by Peter Cocking
Text design by Ingrid Paulson
Cover photograph by Sian Irvine/Dorling Kindersley/Getty Images
Maps by Eric Leinberger
Pipeline map information courtesy of Petr Cizek

Printed and bound in Canada by Friesens
Printed on acid-free paper that is forest friendly (100% post-consumer recycled paper)
and has been processed chlorine free
Distributed in the U.S. by Publishers Group West

We gratefully acknowledge the financial support of the Canada Council for the Arts,
the British Columbia Arts Council, the Province of British Columbia through
the Book Publishing Tax Credit, and the Government of Canada through the Book Publishing
Industry Development Program (BPIDP) for our publishing activities.

NOTE: All gallon measurements in this book are in imperial gallons.
One imperial gallon is approximately 1⅕ U.S. gallons.

To the citizens of Alberta

"Our present 'leaders'—the people of wealth and power—
do not know what it means to take a place seriously:
to think it worthy, for its own sake, of love and study and careful
work. They cannot take any place seriously because they must
be ready at any moment, by the terms of power and wealth
in the modern world, to destroy any place."

WENDELL BERRY, "OUT OF YOUR CAR, OFF YOUR HORSE,"

ATLANTIC MONTHLY, FEBRUARY 1991

CONTENTS

．．．．．．．．．．．．

DECLARATION OF
A POLITICAL EMERGENCY

....

I The world's oil party is coming to a dramatic close, and Canada
has adopted a new geodestiny: providing the United States with
bitumen, a low-quality, high-cost substitute.

II Northern Alberta's bituminous sands, a national treasure, are the
globe's last great remaining oil field. This strategic boreal resource
has attracted nearly 60 per cent of all global oil investments.
Every major multinational and nationally owned oil company has
staked a claim in the tar sands.

III Neither Canada nor Alberta has a rational plan for the tar sands
other than full-scale liquidation. Although the tar sands could
fund Canada's transition to a low-carbon economy, government has
surrendered the fate of the resource to irrational global demands.
At forecast rates of production, the richest deposits of bitumen
will be exhausted in forty years.

IV Nations become what they produce. Bitumen, the new national
staple, is redefining the character and destiny of Canada. Rapid
development of the tar sands has created a foreign policy that

favours the export of bitumen to the United States and lax immigration standards that champion the import of global bitumen workers. Inadequate environmental rules and monitoring have allowed unsustainable mining to accelerate. Feeble fiscal regimes have enriched multinationals and given Canada a petrodollar that hides the inflationary pressures of peak oil. Canada now calls itself an "emerging energy superpower." In reality, it is nothing more than a Third World energy supermarket.

v Investment in the tar sands, including pipelines and upgraders, now totals approximately $200 billion.* The tar sands boom has become the world's largest energy project, the world's largest construction project, and the world's largest capital project. No comprehensive assessment of the megaproject's environmental, economic, or social impact has been done.

vi Thanks to rapid tar sands development, Canada now produces more oil than Texas or Kuwait. Since 2001, Canada has surpassed Saudi Arabia as the largest single exporter of oil to the United States. Canadian crude now accounts for nearly one-fifth of all U.S. oil imports. If development continues unabated, Canada will soon provide the fading U.S. empire with nearly a third of its oil, while half of Canada's own citizens remain dependent on insecure supplies from the Middle East.

vii Rapid tar sands development has become a central goal of the Security and Prosperity Partnership of North America (SPP), an elite plan to create a North American economic union. U.S. energy policy openly advocates for more pipelines and transmission lines to ease growing shortages in energy supply for U.S. citizens, who currently consume 25 per cent of the world's oil. Representatives from the Mexican government attended meetings in 2006 in

---

* Dollar amounts throughout are given in Canadian dollars.

Houston, Texas, about rapid tar sands development. Rapid energy integration will inescapably lead to political integration in a North American union dominated by the United States.

VIII Bitumen is a signature of peak oil and a reminder, as every beer drinker knows, that the glass starts full and ends empty. Half of the world's cheapest and cleanest oil has been consumed. The reality of depletion now demands the mining of the dirtiest. It takes the excavation of two tons of earth and sand to make one barrel of bitumen.

IX Each barrel of bitumen produces three times as much greenhouse gas as a barrel of conventional oil. The tar sands explain why the Canadian government has spent more than $6 billion on climate-change programs for the last fifteen years and met not one target.

X Bitumen is one of the world's most water-intensive oil products. Each barrel requires the consumption of three barrels of fresh water from the Athabasca River, which is part of the world's third-largest watershed. Every day, Canada exports one million barrels of bitumen to the United States and three million barrels of virtual water.

XI Industry in the tar sands uses as much water every year as a city of two million people. Ninety per cent of this water ends up in the world's largest impoundments of toxic waste: the tailings ponds. Industrial water monitoring on the Athabasca River is a fraud. Canada has no national water policy and one of the worst records of pollution enforcement of any industrial nation.

XII The tailings ponds, located along the Athabasca River, leak or seep into groundwater. For the last decade, the downstream community of Fort Chipewyan has documented rare cancers.

XIII To mine or steam out bitumen, the tar sands industry burns enough natural gas every day to heat four million homes. At this rate of consumption, the project could severely compromise the nation's natural gas supplies by 2030.

XIV The rapid depletion of natural gas in the tar sands is driving Canada's so-called nuclear renaissance. Canada may well become the first nation to use nuclear energy not to retire fossil fuels but to accelerate their exploitation.

XV Bitumen development will never be sustainable. The megaproject will eventually destroy or industrialize a forest the size of Florida and diminish the biological diversity and hydrology of the region forever.

XVI Oil hinders democracy and corrupts the political process through the absence of transparent reporting and clear fiscal accounting. Alberta, a classic petrostate, has one of the least accountable governments in Canada as well as the lowest voter turnout.

XVII Without long-term planning and policies, Canada and Alberta will fail to secure reliable energy supplies for Canadians, to develop alternative energy sources for the country, or to create valuable resource funds for the future. Unlike the governments of Norway and Alaska, the government of Canada stands to leave its citizens a singular legacy of exponential neglect and watershed destruction.

XVIII A business-as-usual case for the tar sands will change Canada forever. It will enrich a few powerful companies, hollow out the economy, destroy the world's third-largest watershed, industrialize nearly one-quarter of Alberta's landscape, consume the last of the nation's natural gas supplies, and erode Canadian sovereignty.

xix The destructiveness of the tar sands is not inevitable. But Canadians and Albertans have become too tolerant of the politicians who compromise the nation's energy security as well as the next generation's future. Instead of liquidating the tar sands for global interests, Canada can use the resource for transition to a low-carbon economy.

xx Every Canadian who drives a car is part of this political emergency. And every Canadian can be part of the solution.

xxi The real work of transforming Canada's fossil fuel–dependent economy will not be big and glamorous. It will be humbling, yet rewarding. Our tasks, as social critic Wendell Berry has noted, "will be too many to count, too many to report, too many to be publicly noticed or rewarded, too small to make anyone rich or famous."

xxii We must begin today.

ONE

CANADA'S GREAT RESERVE

....

"Canada, so near and friendly a neighbor that her resources
cannot be thoughtfully considered foreign and alien, has a vast bed
of tar sands in Alberta, the largest known deposit of oil in the world."

SATURDAY EVENING POST, 1943

LONG BEFORE FORT McMurray's video-gambling halls lured Croatian welders, Chinese labourers, Venezuelan engineers, and American whores, Charles Mair saw the global boom foretold, as perhaps only a poet can. During his two-thousand-mile trek through the southern half of the Mackenzie River Basin more than a century ago, a "Go North" future resolutely declared itself to Mair.

The Protestant wordsmith, fur trader, and fervent nationalist coined the now-outdated expression "Canada First." He wrote a play about the great war chief Tecumseh and the temptations of American materialism. As a champion of the beneficence of the British Empire, Mair opposed the Northwest Rebellion but defended the rights of Aboriginal people. He was, in short, the Rudyard Kipling of a Canada that no longer exists.

In 1899, Mair accompanied fifty members of the Treaty Eight and the Half-breed Scrip commissions deep into the forests of the

Unorganized District of Athabasca. He saw the region as "a *terra incognita* — rude and dangerous." Mair understood that the still fur-rich country had no allurements for the average citizen beyond some iconic stories about "barbarous Indians and perpetual frost." Those stories were why he went.

As one of the treaty commission's secretaries, Mair understood the government's real interest in the region. In the late 1880s, a federal report on the "inexhaustible" tar sands called them "the most extensive petroleum field in America, if not the world" and predicted that they would soon rank "among the chief assets comprised in the Crown Domain of the Dominion." A region that rich in economic resources, reasoned Mair, had to be "placed by treaty at the disposal of the Canadian people." The Klondike Gold Rush had brought a noisy gang of U.S. treasure seekers into "Canada's Great Reserve," and that worried Ottawa. It also goaded the government's treaty-making machinery into action. The commission sought to secure access to the oil reserve by recognizing Aboriginal claims with a few dollars and 160 acres of land for each "Red Brother." As with most North American treaties, Treaty Eight aimed to transform forest nomads and fishers into immigrant farmers.

Mair travelled nearly the breadth of the Peace River and much of the Athabasca River with Métis trackers, Catholic missionaries, and members of the North West Mounted Police. During treaty signings, he whimsically recorded the poetic names of Cree men and women, names that suggested the uniqueness of the place: One in the Skies, The Man Who Stands with the Red Hair, Listener to the Unseen Rapids, She Sits in Heaven, Grand Bastard.

One evening as Mair drifted by boat down the Peace River, he beheld the northern lights, what the Cree still call Dance of the Spirits. As did the Cree, he knew that these bluish-green lights heralded great change. After days of vigorous rowing, Mair's party passed through a land "begirt with aspens" and eventually entered Lake Athabasca (The Lake of the Marsh), where lay "the most extensive marshes and feeding-grounds for game in all Canada." It remains one of the largest staging areas for waterfowl in the world.

On the great lake Mair camped with his commission mates at Fort Chipewyan, the sheltered capital of Canada's fur trade and Alberta's oldest settler community. Here, explorers as famous as Alexander Mackenzie and as infamous as Peter Pond, a serial murderer, sat down to fabled feasts of beaver, pickerel, and bison after weeks of starving in the bush. At one time the busy trading post boasted the best library in all of northern Canada. Mair observed that the Chipewyan of Lake Athabasca, a Dene people, spoke the same language as the Apache and dressed plainly. They also held to some fantastic beliefs, namely "that the mastodon still exists in the fastnesses of the Upper Mackenzie" and that this monster was several times bigger than a buffalo.

From the fort Mair boarded the steamer *Grahame* to travel down the mighty Athabasca in the company of "120 baffled Klondikers." The Hudson's Bay Company had given the Americans free passage home to get rid of them. In their gold lust, the "marauders" had killed horses, stolen dogs, smashed bear traps, and raided one Aboriginal village after another throughout the north. Mair took a startling photograph of a solitary woman who had abandoned her "duffer" husband. The confident adventurer serenely wore a bowie knife and a revolver.

One morning, as the steamer rounded a bend in the river, the Klondikers spied three swimming moose. The Americans ecstatically popped off hundreds of volleys of lead. Eventually, they hauled aboard a well-killed three-year-old bull that was "bled and flayed" and served for dinner. At a place called Poplar Point, Mair watched as the steamer stopped to take aboard familiar northern cargo: a white man's corpse "completely enclosed in a transparent coffin of ice."

But Mair didn't see the grand and impossible future of Canada until the steamer docked at Fort McMurray, a "tumble-down cabin and trading-store." That's where he encountered the impressive tar sands, what Alexander Mackenzie had described as "bituminous fountains" in 1778 and what federal botanist John Macoun almost a century later called "the ooze." Federal surveyor Robert Bell described an "enormous quantity of asphalt or thickened petroleum" in 1882. Mair called the tar sands simply "the most interesting region in all the North."

The tar was everywhere. It leached from cliffs and broke through the forest floor. Mair observed giant clay escarpments "streaked with oozing tar" and smelling "like an old ship." Wherever he scraped the bank of the river, it slowly filled with "tar mingled with sand." The Cree told him that they boiled the stuff to gum and repair canoes. One night Mair's party burned the tar like coal in a campfire.

In his now largely forgotten narrative *Through the Mackenzie Basin,* Mair included a prophetic paragraph about "Nature's chemistry," as he called it: "That this region is stored with a substance of great economic value is beyond all doubt, and, when the hour of development comes, it will, I believe, prove to be one of the wonders of Northern Canada."

That hour is now upon us. The region's fame has spread to France, China, South Korea, Saudi Arabia, and Norway. According to the Alberta government, the vast storehouse of the tar sands will solve the world's energy shortages. But greed and moral carelessness have turned the wonder of Canada's Great Reserve to dread.

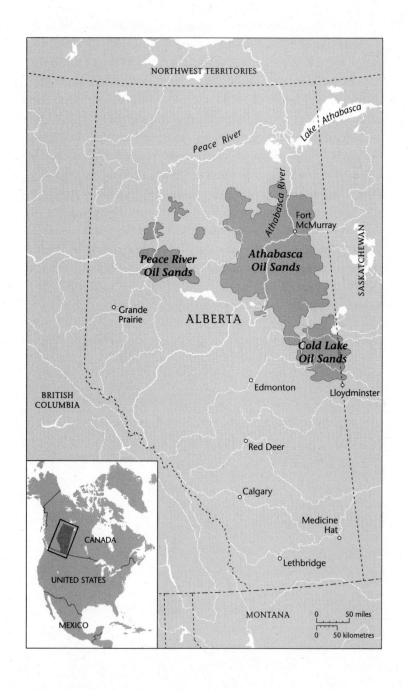

NORTHWEST TERRITORIES

Peace River

Lake Athabasca

Athabasca River

Fort McMurray

Peace River Oil Sands

Athabasca Oil Sands

SASKATCHEWAN

○ Grande Prairie

ALBERTA

Cold Lake Oil Sands

BRITISH COLUMBIA

○ Edmonton

Lloydminster

○ Red Deer

○ Calgary

Medicine Hat ○

○ Lethbridge

CANADA

UNITED STATES

MEXICO

MONTANA

0   50 miles

0   50 kilometres

TWO

IT AIN'T OIL

....

"I do not think there is any use trying to make out
that the tar sands are other than a 'second line of defense'
against dwindling oil supplies."

KARL A. CLARK, RESEARCH ENGINEER, LETTER TO OTTAWA, 1947

A COUPLE OF years ago, while most Canadians were out shovelling the
sidewalk or sipping a coffee at Tim Hortons, the nation quietly became
a petrostate. Prime Minister Stephen Harper, the son of an Imperial
Oil executive, made the startling announcement in several world capi-
tals. Harper pronounced Canada "an emerging energy superpower" and
a Saudi Arabia in the making. Canada was the only non–OPEC country
"with growing oil deliverability," he boasted. "We are a stable, reliable
producer in a volatile, unpredictable world."

The driver of Canada's conversion to petrodollars is a dirty resource
called bitumen, what Harper described as an "ocean of oil-soaked sand."
This unconventional oil now shapes the Canadian economy and politics
the same way the fur trade once did. Three hundred years ago, Canada
supplied Europe's fashion industry with beaver pelts. Today it pipes unre-
fined bitumen to U.S. refineries to keep that country's sputtering
economy supplied with oil.

Lots of countries can claim deposits of oil-soaked sand, but none is as great in size as Canada's. (Venezuela's Orinoco River Basin is a close second.) The main deposit lies in northeastern Alberta, close to the Saskatchewan border. Sizable reserves also lie under the boreal forest near the cities of Cold Lake and Peace River.

Like most hydrocarbon formations, the tar sands are a fixed inheritance, the product of ancient marine life (largely sun-baked algae and plankton). Some 200 million to 300 million years ago, geological forces started to compress and cook the dead plants and creatures, then degraded the remaining mess with bacterial activity. Good cooking results in light oil. Bad cooking makes bitumen. Fifty per cent of Canada now depends on a half-baked fuel synthesized by plants and stored as chemical energy at a time when dinosaurs briefly ruled.

INDUSTRY EXECUTIVES AND many Canadian politicians get upset when they hear the term *tar sands*. They think tar is "greenie speak," a tasteless pejorative for the largest deposit of oil outside of Saudi Arabia (175 billion barrels). Marketers and CEOs prefer the word *oil* in relation to the sands because it sounds abundant, accessible, and clean. *Oil* raises investment cash faster than *tar* does, and it reassures consumers nervous about the ever-rising price of fossil fuels. An emerging energy superpower such as Canada doesn't mine and upgrade nasty bitumen; it produces oil. The Alberta government says it makes sense to describe the resource as oil sands "because oil is what is finally derived from bitumen." If that lazy reasoning made sense, Canadians would call every tomato *ketchup* and every tree *lumber*. Passing off tarlike bitumen as oil is about as accurate as calling an aspen tree a Douglas fir, or a donkey a horse.

Bitumen, a thick, heavy substance, looks and smells like tar or black molasses. It makes a competent road cover, and the Hudson's Bay Company used it to repair leaky roofs and canoes. But it's a poor man's hydrocarbon and one very dirty oil product.

Dr. Steven Kuznicki, a scholar at the Imperial Oil–Alberta Ingenuity Centre for Oil Sands Innovation, calls bitumen some of the "ugliest stuff you ever saw...contaminated, non-homogenous and ill-defined...

Bitumen is five per cent sulphur, half a per cent nitrogen and 1,000 parts per million heavy metals. Its viscosity [stickiness] is like tar on a cold day. That's ugly."

Bitumen contains more carbon than light crude does. In fact, it is so hydrogen deficient that it requires either the removal of carbon atoms or the addition of hydrogen atoms to be of much use at all.

Rick George, president and CEO of Suncor, highlighted bitumen's thoroughly roguish nature in a 2008 speech to the World Heavy Oil Congress. Although he prefers the term *oil sands,* George unwittingly made a good argument for calling the stuff *tar.* Bitumen may contain a hydrocarbon, he said, but you can't use it as a lubricant because "it contains minerals nearly as abrasive as diamonds." You can't pump it, because "it's as hard as a hockey puck in its natural state." It doesn't burn all that well, either; "countless forest fires over the millennia have failed to ignite it." All in all, George said, the resource "has no value, other than the value we create."

Bitumen can't be sucked out of the ground like Saudi Arabia's black gold. It took an oddball combination of federal and provincial scientists and American entrepreneurs nearly seventy years from the time of Mair's visit to the tar sands (and billions of Canadian tax dollars) to figure out how to separate bitumen from sand. They finally arrived at a novel solution: brute force.

Extracting bitumen from the forest floor is done in two earth-destroying ways. About 20 per cent of the tar sands are shallow enough to be mined by three-storey-high, four-hundred-ton Caterpillar trucks from Illinois (drivers compare the experience to navigating an apartment building) and $15-million Bucyrus electric shovels from Wisconsin.

The open-pit mining operations look more hellish than an Appalachian coal field. To coax just one barrel of bitumen from the Athabasca sand pudding, companies must mow down hundreds of trees, roll up acres of soil, drain wetlands, dig up four tons of earth to secure two tons of bituminous sand, and then give those two tons a hot wash. The process costs approximately $100,000 per flowing barrel, making bitumen one of the planet's most expensive fossil fuels. Every other day, the open-pit mines move enough dirt and sand to fill Yankee Stadium or Toronto's Rogers Centre. Since 1967, one major mining company has moved enough earth (2 billion tons) to build seven Panama canals. Larry Burns, vice-president of research and development and strategic planning for General Motors, characterizes this kind of unconventional earth moving as damned messy: "Does it make sense for Alberta to be creating an oilsands industry? . . . I think it's a pretty bizarre way to get gasoline to a corner station. It's an awful lot of capital and an awful lot of work to pull it off." In the prestigious journal *Petroleum Economist,* one anonymous tar sands executive described the open-pit mines as an "environmental freak show."

Most of the tar sands, however, lie in such deep formations that the bitumen must be steamed or melted out of the ground, with the help of a bewildering array of pumps, pipes, and horizontal wells. Engineers call the process in situ (in place). The most popular in situ technology is steam-assisted gravity drainage (SAGD).

"Think of a big block of wax the size of a building," SAGD expert Neil Edmunds explains. "Then take a steam hose and tunnel your way in and melt all the wax above. It will drain to the bottom where it can be collected. That's what SAGD does to bitumen."

SAGD can also kill the living heart of a forest. A typical project occupies a three-mile by three-mile area and destroys 7 per cent of the land. But the technology's supporting roads, pipelines, and seismic lines industrialize the forest so irresolutely that it makes the land inhospitable for much wildlife. A 2008 report by the industry-funded Cumulative Environmental Management Association disclosed that SAGD, as currently designed, would extirpate caribou, fish, bear, and moose over a region ranging from one to three million acres in size. Even better industrial practices don't make much of a difference. "They don't change the destination for wildlife. With SAGD there is a profound loss of species, though [greener practices] do take longer to wipe them out," explains study author Brad Stelfox.

SAGD technology burns enough natural gas, for boiling water into steam, to heat four million North American homes every day. In fact, natural gas now accounts for more than 60 per cent of the operating costs for a SAGD project. Using natural gas to melt a resource as dirty as bitumen is, as one executive said, like "burning a Picasso for heat." Dr. Eddy Isaacs of the Alberta Energy Research Institute calculates that "the amount of energy required to produce a barrel of synthetic crude oil is about a third of the energy in a barrel of bitumen."

Working harder to create less energy marks bitumen as a third-rate product. On average, it takes one barrel of oil, or its energy equivalent, to pump out anywhere between twenty and sixty barrels of cheap oil. In contrast, the U.S. Department of Energy calculates that an investment of one barrel of energy yields between four and five barrels of bitumen from the tar sands. Some experts figure that the returns on energy invested may be as low as two or three barrels. Shell geophysicist Marion King Hubbert predicted that a society based on fossil fuels will come to a dead end "when the energy cost of recovering a barrel of oil becomes greater than the energy content of the oil." But neither the Alberta nor the Canadian government has done a thorough energy accounting yet.

Bitumen's low-energy returns and earth-destroying production methods explain why the unruly resource can cost nearly twenty times more than conventional crude to produce and upgrade, at $100,000 per flowing barrel. Given its impurities, bitumen often sells for half the price of West Texas crude. (Much to everyone's surprise, peak oil narrowed the price differential to 20 per cent in 2008.)

Only complex refineries can add value to highly variable grades of bitumen. As a result, the resource routinely supports a 63 per cent greater price volatility than does conventional oil. To date, no transparent pricing framework exists for bitumen. In 2007, Alberta Energy admitted that "the markets which Alberta crude can currently access do not have sufficient heavy oil conversion capacity to always ensure good prices for bitumen."

In 1983, engineer Donald Towson made a good case for calling the resource tar, not oil, in the *Encyclopedia of Chemical Technology*. He argued that the word accurately captures the resource's unorthodox makeup, which means it is "not recoverable in its natural state through a well by ordinary production methods." Towson noted that bitumen not only has to be diluted with light oil to be pumped through a pipeline but requires a lot more processing than normal oil. (Light oil shortages are so chronic that industry imported 50,000 barrels by rail last year to the tar sands.) Even after being upgraded into "synthetic crude," the product requires more pollution-rich refining before it can become jet fuel or gasoline.

"Professor Nositall" at the Oil Sands Discovery Centre in Fort McMurray accepts these facts. The cartoon character reminds visiting children and journalists that untreated bitumen will not flow an inch: "It is too thick to pump, too thick to collect in wells and too thick to move in pipelines."

Bitumen is what a desperate civilization mines after it's depleted its cheap oil. It's a bottom-of-the-barrel resource, a signal that business as usual in the oil patch has ended. To use a drug analogy, bitumen is the equivalent of scoring heroin cut with sugar, starch, powdered milk, quinine, and strychnine. Calling the world's dirtiest hydrocarbon "oil" grossly diminishes the resource's huge environmental footprint. It also distracts North Americans from two stark realities: we are running out of cheap oil, and seventeen million North Americans run their cars on an upgraded version of the smelly adhesive used by Babylonians to cement the Tower of Babel.

That ancient megaproject did not end well. Without a disciplined plan for them, the tar sands won't either.

THREE

THE VISION OF HERMAN KAHN

....

"For America, buying oil from Canada should be akin to
buying a used pickup truck from your brother-in-law...you need
the truck, he needs the money, and you are each pretty sure
that the other is not seeking your total annihilation."

FIRST COMMENTARY, 2003

THE MEDIA CALLED him Dr. Megadeath, and rightly so. Herman Kahn
worked for the famous RAND Corporation, where he thought a lot about
the unthinkable: nuclear war. As a military strategist and systems the-
orist in the early 1960s, Kahn argued that a thermonuclear war might
destroy the ozone or start an ice age, but it would not end the human
race—in part, he said, because "war is a terrible thing, but so is peace.
The difference seems to be a quantitative one of degree and standards."
After making the prospect of enduring a nuclear war a matter of comic
relief (the affable pundit served as one of the real-life models for Stan-
ley Kubrick's film *Dr. Strangelove*), Kahn moved on to the business of
predicting the future at the Hudson Institute, a U.S. think tank he
helped to found.

From the institute's Washington, D.C., headquarters, the rotund
technocrat and eternal optimist, a kind of Zero Mostel of futurology,

invariably predicted that humans, no matter what the challenge, would come out smiling. Kahn saw no limits to growth: only brimming cornucopias. For the benefit of short-sighted governments and politicians alike, he pioneered the art of the scenario, "one way to force oneself and others to plunge into the unfamiliar." In 1973 he proposed that Canada take such a plunge into the tar sands.

At the time, Middle East oil barons had turned off the tap. American motorists were lining up at the pump, and conventional U.S. oil production, as had been famously predicted by Shell geophysicist Marion King Hubbert, was in steep decline. The United States, once the world's number-one oil supplier, now had to buy a third of its oil from abroad. The vulnerability of a superindustrial society dependent on fuel from the Persian Gulf worried even cheerful global thinkers.

Kahn, who knew his geography as well as his geology, peered into the present to notice that the tar sands lay in friendly and largely unexploited territory. The sands were secure, they were vast, and they required no exploration costs. More important, the United States had already staked out a big claim in Canada's Great Reserve. Against all economic odds, visionary J. Howard Pew, then the president of Sun Oil and the seventh-richest man in the United States, had built a mine and an upgrader (now Suncor) on the banks of the Athabasca River in 1967. Pew's folly, then the largest private development ever built in Canada, would lose money for twenty years by producing the world's most expensive oil at more than $30 a barrel. But Pew reasoned that "no nation can long be secure in this atomic age unless it be amply supplied with petroleum." Given the inevitable depletion of cheap oil, he recognized that the future of North America's energy supplies lay in expensive bitumen. (A consortium of four U.S.–owned major oil companies, including Imperial Oil, later seconded Pew's assessment by building Syncrude next door to Suncor. The company once described its 170,000-acre lease site on the Athabasca River by bragging that "25 per cent of the world's countries are smaller than that.")

Kahn, who chatted regularly with folks such as Henry Kissinger, U.S. President Richard Nixon's national security advisor, also appreciated the subtleties of Project Independence, the title given to U.S. government energy policy in the early 1970s. The policy stated that "there is an advantage to moving early and rapidly to develop tar sands production" because it "would contribute to the availability of secure North American oil supplies." Mining Canada's forest for bitumen would give the United States some time to figure out how to economically exploit its own dirty oil in places such as Colorado's oil shales and Utah's tar sands.

In the fall of 1973, Kahn and Montreal associate and economist Marie-Josée Drouin (she is now married to hedge fund manager Henry Kravis and serves as a senior fellow at the Hudson Institute) flew off to Ottawa. There they made an incredible offer to the Canadian government. Given the current energy crisis and OPEC's reluctance to boost oil production, Kahn hailed the bituminous sands of northern Alberta as a global godsend. He then presented a tar sands crash-development program to Prime Minister Pierre Elliott Trudeau and Energy Minister Donald Macdonald.

Like everything about Kahn, his rapid development scheme was big and bold. (A crash program, said Kahn, was really "overnight go-ahead decision making.") This one called for the construction of twenty gigantic open-pit mines with upgraders on the scale of Syncrude, soon to be one of the world's largest open-pit mines. The futurist calculated that the tar sands could eventually pump out two million to three million barrels of oil a day, all for export. Canada wouldn't have to spend a dime, either. A global consortium formed by the governments of Japan, the United States, and some European countries would put up the cash: a cool $20 billion. Korea would provide thirty thousand to forty thousand temporary workers, who would pay dues and contribute to pension plans to keep the local unions happy. Kahn pointed out that Canada would receive ample benefits: the full development of an underexploited resource, high revenues, a refining industry, a secure market, and lots of international trade.

The audacity of the vision stunned journalist Clair Balfour at the *Financial Post*, who wrote, "It would be as though the 10,000 square miles of oil sands were declared international territory, for the international benefit of virtually every nation but Canada."

Biologists and ecologists understood that the environmental consequences of digging up a forest in a river basin that contained one-fifth of Canada's fresh water could be enormous. According to Larry Pratt's lively account of Kahn's presentation in his book *The Tar Sands*, one federal government official calculated that the megaproject would dump up to twenty thousand tons of bitumen into the Athabasca River every day and destroy the entire Mackenzie basin all the way to Tuktoyaktuk. Studies and reports completed in 1972 had warned that the construction of "multi-plant operations" would "turn the Fort McMurray area of northeastern Alberta into a disaster region resembling a lunar landscape" or a "biologically barren wasteland."

Kahn, who opposed "emotional sloppy thinking," dismissed the environmental sentimentalists. Why not simply sacrifice the largely uninhabited wilderness for global energy security? He argued that northern Alberta was "a relatively undesirable environment anyway. Its restoration would not be a matter of aesthetic quality." Because the oil shales in Colorado lay in scenic Rocky Mountain country, Kahn thought it made more sense to dig up the mosquito-infested, muskeg-laden boreal forest first. After hearing Kahn's impassioned pitch, one Canadian economist mused, "I suppose if one were at war, it's surprising the things one would do and things you would ignore . . . If you were at war, you'd use up the Athabasca River and say the hell with it."

The Canadian government briefly entertained Kahn's scenario but then declined, citing largely economic reasons. The U.S. government persisted, offering an $8-billion industrial assistance package to kick off Kahn's crash program. Even after mulling that sum over, Ottawa concluded that Kahn's megascheme could overheat the economy, create steel shortages, unsettle the labour market, drive up the value of the Canadian dollar, and generally change the nation beyond recognition. The tar sands would also be needed to meet future domestic energy needs. "I don't know, within the world community, why we should feel any obligations to rush into such large-scale production, rather than leave it in the ground for future generations," reasoned Donald Macdonald.

Kahn parked his vision for a while. Several years later, he advocated that the American government invest $20 billion in U.S. oil shales (ugly rock impregnated with oil) to build as many as ten fifty-thousand-barrel-a-day plants "to reduce U.S. dependence on OPEC." But the energy shock eventually passed. People briefly conserved, oil prices dropped, and OPEC got smarter. The United States quietly turned to Mexico for oil and worked out a "special relationship" with Saudi Arabia. But Kahn never gave up on his big idea. He knew that "the long-term threat of ever rising petroleum prices" wouldn't go away. Thirty years later, events proved him right.

WHAT IS ARGUABLY the world's last great oil rush is taking place today on a scale that would have stunned even the unflappable Kahn. Instead of a conservative twenty projects over a decade, Alberta has approved nearly one hundred mining and in situ projects. That makes the tar sands the largest energy project in the world, bar none.

The size of the resource being exploited has grown exponentially. The bitumen-producing zone contains nearly 175 billion barrels in proven reserves, which makes it the single-largest pile of hydrocarbons outside of Saudi Arabia. At 54,000 square miles, the zone covers a forest region five times larger than the 10,000 square miles Kahn originally slated for global demolition. Alberta Energy proudly reports that the landscape being industrialized by rapid tar sands development could easily accommodate one Florida, two New Brunswicks, four Vancouver Islands, or twenty-six Prince Edward Islands. That's nearly a quarter of the landmass of Alberta.

The collective value of the current crash program makes Kahn's original dollar figures look positively conservative. Instead of $20 billion, tar sand investments now total nearly $200 billion. That hard-to-imagine sum easily makes the tar sands the world's largest capital project. True to Kahn's prediction, the money comes from around the globe, including France, Norway, China, Japan, and the Middle East. But approximately 60 per cent of the cash hails from south of the border. Although a Korean contingent of cheap labour has yet to appear on the horizon, an itinerant army of bush workers from China, Mexico, Hungary, India, Romania, and Atlantic Canada, among other places, is now digging away. Every day, companies decry the shortage of labour by advocating for looser immigration laws.

The Alberta tar sands are a global concern. The Abu Dhabi National Energy Company (TAQA), an expert in low-cost conventional oil production, bought a $2-billion chunk of bitumen real estate just to be closer to the world's largest oil consumer, the United States. South Korea's national oil company owns a piece of the resource, as does Norway's giant national oil company, Statoil, which just invested $2 billion. Total, the world's fourth-largest integrated oil and gas company, with operations in more than 130 countries, plans to steam out two billion barrels of bitumen. Shell, the global oil baron, lists the Athabasca Oil Sands Project as its number-one global enterprise and plans to produce nearly a million barrels of oil a day—more oil than is produced daily in all of Texas.

Synenco Energy, a subsidiary of Sinopec, the Chinese national oil company, says it will assemble a modular tar sands plant in China, Korea, and Malaysia, then float the whole show down the Mackenzie River. Japan Canada Oil Sands Limited has put up money. India's state oil company is sniffing around, too; as one manager put it, "We are seriously interested in opportunities here." BP, one of the world's largest energy companies, originally refused to join the parade to the tar sands because of the "environmental damage." But BP threw caution aside and teamed up with Husky Oil for a $10-billion project, arguing that "these resources would have to be developed anyway." On and on the list of global oil heavyweights goes.

The current boom makes a mockery of Kahn's original oil production forecasts. As of spring 2008, the tar sands produce 1.3 million barrels a day. (That's more than half of Canada's oil production.) Expansions and new projects will add three million barrels a day to southbound pipelines by 2015. Many forecasts expect the tar sands to reach five million barrels a day by 2030. Incredibly, the Alberta government has even run a Kahn-like scenario to accelerate production to eight million barrels a day by 2050. That would make Alberta another Saudi Arabia.

Kahn's labour forecasts for the oil sands also proved understated. Fifty thousand temporary foreign workers have poured into Alberta to feed the bitumen boom. Bitumen has transformed the federal government's once sleepy Temporary Foreign Worker (TFW) program into a major and controversial tool to secure cheap, disposable labour for 170 "occupations under pressure." For the first time in Canadian history, the number of temporary workers toiling in a province outnumbers the province's legal immigrants. According to the Alberta Federation of Labour, Alberta, as a proportion of its population, "had 12.5 times as many TFWs than the United States" in 2008. (A third of Saudi Arabia's petroleum-driven work force are also temporary guest workers.)

Abuse of guest workers is so widespread that the Alberta government handled 800 complaints in just one three-month period in 2008. Qualified chefs hired in Fiji typically ended up sweeping floors in Alberta, while 120 Chinese construction workers at CNRL's Horizon Mine received only a fraction of their wages due to corrupt contracting practices. Gil McGowan, president of the Alberta Federation of Labour, warned the House of Commons Standing Committee on Citizenship and Immigration in April 2008 that wherever guest-worker programs have been used in Germany or the United States, "they've led to exploitation, the creation of job ghettos and rising social tensions." Yet due to the uncontrolled bitumen boom, Alberta employers applied for 100,000 temporary workers in 2007.

The Trudeau government's fear that Kahn's superbitumen plan would strain the economy has proved bang-on. The nation now boasts a petrodollar that confounds most citizens. Between 2003 and 2006, the Canadian dollar rose from sixty-four cents to eighty-seven cents against the U.S. dollar, nearly parallel with the price of crude oil. Most global currency traders now treat the Canadian dollar as a petrocurrency. Since 2003, Canada's manufacturing sector has been bleeding jobs to oil. Since 2005, energy exports have outperformed all sectors and now account for nearly 10 per cent of Canada's gross domestic product (GDP). Statistics Canada reported in 2008 that annual investment in the tar sands, incredibly, had exceeded spending forecasts for the entire manufacturing base of the country. Meanwhile, the price of steel and concrete has climbed ever upwards. Provinces blessed with hydrocarbons (Alberta, British Columbia, and Saskatchewan) have recorded crazy, Chinese-paced growth, while the economies of Ontario and Quebec have stagnated.

The boom has changed Alberta's social and economic landscape practically beyond recognition. Between 1996 and 2006, more than 700,000 people poured into the province, creating a $7-billion infrastructure shortfall in roads, schools, and hospitals. Labour shortages are so extreme that some tar sands workers commute all the way from Thunder Bay, Ontario, or even St. John's, Newfoundland, every two weeks. The gruelling pace of bitumen's development has also made Alberta a dangerous place to work. In 2007, the boom killed 154 people on the job (a 24 per cent increase over 2006) and injured 34,000. Prosecutions for workplace health and safety violations are as rare as environmental investigations.

To appreciate the scale and economic impact of just one tar sands project, consider Shell's Athabasca Oil Sands Project, forty-seven miles north of Fort McMurray. The complex, which consists of the Muskeg River Mine and Scotford Upgrader, occupies a piece of the boreal forest the size of 33,702 NHL hockey rinks. Some six thousand workers toil at one expansion stage or another. At the Muskeg River Mine, high-school grads earn more than $100,000 a year driving the world's largest trucks (four-hundred-ton vehicles with the horsepower of a hundred pickup trucks) to move $10,000 worth of bitumen a load. The trucks dump the ore into a crusher, which spits the bitumen onto the world's largest conveyor belt, about 1,600 yards long. The bitumen is eventually mixed with expensive light oil and piped to an Edmonton refinery.

Shell's project, one of the world's largest construction gigs, stands as an awe-inspiring testament to the power of industrial consumption.

The boreal-destroying enterprise required 995 miles of pipe and now consumes enough power to light up Burlington, Ontario, a city of 136,000 people. The mine gobbled up enough steel cable to stretch from Calgary to Halifax and poured enough concrete to build thirty-four Calgary Towers. At full production, the plant will kick out 10 per cent of Canada's oil needs. And it's all sustainable, or at least that's what Shell's 2004 Sustainable Development Report says. Neil Camarta, the senior manager who oversaw much of the construction, compared the mine's construction to war and said the whole project took "lots of energy and lots of guts."

Rapid tar sands expansion also takes lots of money. The U.S. government spent $20 billion over a thirteen-year period on the Apollo program to put a man on the moon. To keep approximately seventeen million North Americans in their cars, multinationals have now invested the equivalent of ten Apollo projects in the tar sands. The persistent government failure to stagger or sequence projects (even former Alberta premier Peter Lougheed has advocated for this reasonable measure) has led to astonishing cost overruns, inflationary wages, and incredible waste. The price tag for an open-pit mine plus an upgrader has climbed from $25,000 to between $90,000 and $110,000 per flowing barrel over the last decade. Given that conventional oil requires, on average, $1,000 worth of infrastructure to remove a flowing barrel a day, Houston-based energy investment banker Matthew Simmons, of Simmons & Company International, says that the "energy's pricing committee" has truly flunked in the tar sands. Yet the rising price of oil has largely obscured these extravagant costs. Canadian Prime Minister Stephen Harper told global investors in 2006 that the sands are "an enterprise of epic proportions, akin to the building of the pyramids or China's Great Wall. Only bigger."

A NATION-CHANGING EVENT bigger than China's Great Wall took four significant drivers: U.S. oil demand, a regulator that behaves like a promoter, a government that behaves like an African potentate, and an important document called the *Declaration of Opportunity*.

The seeds for the declaration were sown in the 1990s, a bad time for the tar sands. Dismally low oil prices had depressed the bitumen market. Multinationals shelved a promising $12-billion project as two tar sands pioneers, Suncor and Syncrude, struggled to make ends meet. The average Canadian at the time had no idea that 20 per cent of our oil came from open-pit mines, or what some oil-patch wags called "brute force combined with ignorance." All in all, global investors had written off the tar sands as a black money pit. According to former Syncrude CEO Eric Newell, "The oil sands had just fallen off everyone's radar." (Unlike many current executives in the tar sands, Newell championed quality schools and invested in Aboriginal employment and community infrastructure. He remains one of the province's most respected business leaders.)

In 1993, a group of thirty oil companies and government agencies gathered to raise the declining profile of Canada's Great Reserve. Shortly afterwards, the group formed the National Oil Sands Task Force. It became what Newell calls "the mother of all collaboration." The task force decided to sell the benefits of Canadian self-sufficiency in oil as well as to emphasize the downside of "increased reliance on Middle East oil and politics." It also wanted to beat Venezuela as the world's next great oil prize.

Two years later, the task force released a nifty twenty-five-year strategy, *The Oil Sands: A New Energy Vision.* The report identified the tar sands, which contained a third of the world's known petroleum resources, as "the largest potential private sector investment opportunity for the public good remaining in Western Canada." To entice investors, the plan proposed that the Alberta government reduce a hodgepodge of royalty fees then as high as 30 per cent to a single-digit deal: a generic 1 per cent regime until companies had paid off their multibillion-dollar investments. The plan also advocated that the federal government provide megabreaks in corporate taxes. This new vision rebranded bitumen, the dirty hydrocarbon, as a "knowledge-based, technology-driven, resource of substantial quality and value" as well as "a national treasure."

The task force hoped that the plan would encourage companies to invest $25 billion, create ten thousand jobs, and slowly boost the nation's oil production from 450,000 barrels a day to nearly a million over a twenty-five-year period. It called for logical, staged, and incremental development. "We really sold it," recalls Newell. "It was the most comprehensive lobbying since Free Trade." In 1995, the governments of Canada and Alberta signed on, and the next year the report morphed into a national *Declaration of Opportunity.*

It didn't take long for opportunity to knock. Within two years, U.S. and Canadian oil companies plunked down more than $10 billion for projects. By 1998, the roads in and out of Fort McMurray were humming and the city's hotels were full. Housing prices had jumped by $50,000. Suncor had started to clear-cut an estimated 290,000 trees for its Steep Bank mine, and surveyors and contractors staked out new mine sites for Shell and Syncrude. Bitumen leases that had sold for $6 an acre in 1978 now sold for $120. (By 2006, companies would be paying $486 per acre.)

The speed of growth surprised even executives, who called the tar sands "the real Canadian oil story." Jim Carter of Syncrude crossed his fingers, saying, "We just hope the coming development is logical and managed." Rob Macintosh, director of the Pembina Institute, an energy watchdog, started praying. "The regulatory agencies in Alberta just aren't capable of forecasting, assessing or managing the full environmental effects of proposed expansions in Fort McMurray," he said in 1998. "The public is in the dark."

"Oil sands fever" turned the task force's rational projections into dull paperwork. Announcements for a new mine or pipeline hit the headlines on a monthly basis. Within eight years, "one of the largest industrial expansions in recent Canadian history" surpassed the task force's investment forecast nearly fourfold. Although the strategy had called for a million barrels a day by 2020, the boom delivered that volume nearly sixteen years ahead of schedule. "What was visionary at the time turned out to not be very visionary," says Eric Newell.

The *Declaration of Opportunity* had all kinds of strategic helpers. Former Alberta Premier Ralph Klein often claimed that his government had no plan for the tar sands boom. According to him, it all just happened like magic: "To have a long-range plan would be an interventionist kind of policy which says you either allow them or you don't allow them to proceed. The last thing we want to be is an interventionist government." But Klein's government did intervene, with a heavy hand: it never refused a single tar sands project, thereby accelerating the pace of development.

The Energy Resources Conservation Board (ERCB), the province's oil and gas regulator, became a much more critical driver of rapid tar sands development than "the come and get it fiscal regime" offered by the *Declaration of Opportunity*. The ERCB, founded in 1938, at one time actually said no to projects. But since the 1990s, the politically appointed board has become a captive regulator, largely funded by industry and mostly directed by lawyers and engineers with ties to the oil patch. On paper, the ERCB has a mandate to develop and regulate oil and gas production in the public interest, and it claims to have the world's most stringent rules. But these rules have allowed the board to approve oil wells in lakes and parks, permit sour-gas wells—as poisonous as cyanide—near schools, and endorse the carpet-bombing of the province's most fertile farmland with thousands of coal-bed methane wells and transmission lines. Until recently, the board refused to report the names of oil and gas companies not in compliance with its regulations, citing security reasons. Curiously, the agency has only two mobile air monitors to investigate leaks from 244 sour-gas plants, 573 sweet-gas plants, 12,243 gas batteries, and about 250,000 miles of pipelines. In any given year, the board approves more than 95 per cent of the sixty thousand applications submitted by industry. The ERCB is the kind of institution Herman Kahn was thinking about when he wrote that "a surprising number of government committees will make important decisions on fundamental matters with less attention than each individual would give to buying a suit."

Since the 1996 *Declaration of Opportunity*, the ERCB (often in joint hearings with the Canadian Environmental Assessment Agency) has approved one mining and in situ project after another in the tar sands. The decisions stand as classic examples of bureaucratic neglect and abuse. After hearing in 2006 that the construction of Suncor's $7-billion Voyageur Project would draw down groundwater by three hundred feet, overwhelm housing and health facilities, and result in air quality exceedances for sour gas, benzene, and particulate matter, the board agreed that the project would "further strain public infrastructure" but declared the impacts "acceptable." After the Albian Sands Muskeg River Mine Expansion proposed to dig up 31,000 acres of forest, destroy 170 acres of fish habitat along the Muskeg River, and withdraw enough water from the Athabasca River to fill 22,000 Olympic-sized pools a year, the board concluded in 2006 that the megaproject was "unlikely to result in significant adverse environmental effects." In 2007, when Imperial Oil's Kearl project proposed to plant four open-pit mines in a seventy-seven-square-mile area, producing more greenhouse gas emissions than 800,000 passenger vehicles in Canada, the board repeated its favourite cliché: the project "is not likely to cause significant adverse environmental effects."

The ERCB, which neglected to open an office in Fort McMurray until 2003, has to date given a thumbs-up to more than one hundred tar sands projects, both big and small. Basically, anyone who wants to build a plant gets one in the province. "Alberta has been very oil friendly and very development friendly and no project has been rejected outright," reported Robert Mason, vice-president of TD Securities at the Western Canada Oil Sands Summit in 2004.

Maurice Dusseault, a global expert on unconventional oil and a widely published academic, accurately described the board's character in a 2002 report for Alberta Energy on heavy oil developments. Said Dusseault: "There is a general consensus in Alberta (except within oil companies) that the environmental aspects of the oil and gas industry require better management and enforcement by the ERCB, or alternatively that the authority should reside with Alberta Environment, a separate provincial department." Dusseault questioned whether ERCB's dual mandate of approving projects and managing royalties was a wise one: "Giving the agency responsible for production and royalties the mandate to also enforce regulations leads to a difficult internal conflict of interest. The result is usually a clashing of goals. Most commonly it is the enforcement of environmental regulations that suffers."

THE WORLD'S LARGEST energy consumer has driven the pace and scale of the bitumen rush as actively as have Alberta's rubber-stamp regulators. With just 5 per cent of the world's population, the United States now burns up 20.6 million barrels of oil a day, or 25 per cent of the world's oil supply. Thanks to bad planning and an aversion to conservation, the empire must import two-thirds of its liquid fuels from foreign suppliers, often hostile ones. "The reality is that at least one supertanker must arrive at a U.S. port every four hours," notes Swedish energy expert Kjell Aleklett. "Any interruption in this pattern is a threat to the American economy." This crippling addiction has increasingly become an unsustainable wealth drainer. In 2000, the United States imported $200 billion worth of oil, thereby enriching many of the powers that seek to undermine the country. By 2008, it was paying out a record $440 billion annually for its oil.

The undeclared crash program in the tar sands has transformed Canada's role in the strategic universe of oil. By 1999, the megaproject had made Canada the largest foreign supplier of oil to the United States. By 2002, Canada had officially replaced Saudi Arabia and Mexico as America's number-one oil source, an event of revolutionary significance. Canada currently accounts for 18 per cent of U.S. oil imports (that's 12 per cent of American consumption), and the continuing development of the tar sands will double or triple those figures. Incredibly, only two in ten Americans and three in ten Canadians can accurately identify the country that now keeps the U.S. economy tanked up.

U.S. Vice President Dick Cheney, a devoted oil man, was among the first to notice all the frantic digging in northern Alberta. His 2001 National Energy Policy, a controversial document drawn up largely in secret by oil executives, declared that the United States was facing "its most serious energy shortages since the oil embargoes of the 1970s."

Predictably, the policy identified the exploration, production, and consumption of more oil and gas as the best solution. Cheney highlighted the tar sands as "a pillar of sustained North American energy and economic security." He also called for a "North American Energy Framework to expand and accelerate cross border energy investment in oil and gas pipelines." A month after Cheney put forward his policy, the energy ministers of Mexico, Canada, and the United States formed the North American Energy Working Group "to enhance North American energy trade and interconnections consistent with the goal of sustainable development." (Nine different energy groups are now making plans to unite the continent's oil, gas, and electricity markets.)

After 9/11 and the falling Twin Towers, proposals for accelerated continental energy integration flowed across the border faster than new ERCB approvals in the tar sands. The goal was no longer energy independence but "interdependence." In 2003, U.S. Ambassador to Canada Paul Cellucci declared that the time had come "to complete the integration of our energy markets." The following year the Canadian Council of Chief Executives, led by Rick George, the Colorado-born CEO of Suncor, championed a resource security pact and regulatory convergence because "Canada has a critical role to play in ensuring the energy security of the continent."

With Kahn-like flair, one think tank after another has mined the same tarry concept. At the 2004 North American Forum on Integration in Monterrey, Mexico, the Mexican oil analyst Lourdes Melgar argued that North American Free Trade Agreement (NAFTA) partners should also be energy partners because "the sophisticated and dynamic North American economy demands a continuous and rising supply of energy." The Council of the Americas' Energy Action Group later added that an integrated energy marketplace would naturally require "the standardization of regional and subregional laws, taxes, royalties and transmission rates."

In 2005, rapid tar sands development achieved top billing in a bold proposal by wealthy businessmen and government elites for the political and economic integration of the continent. Their report, *Building a North American Community,* outlined a scenario in which Mexico would supply the cheap labour and Canada the cheap energy for a U.S.–dominated North American economy. The report gushed that "Canada's vast oilsands, once a high cost experimental means of extracting oil, now provide a viable new source of energy that is attracting a steady stream of multi-billion dollar investments and interest from countries such as China and they have catapulted Canada into second place in the world in terms of proved oil reserves." Blessed with this sort of abundance, the report advised, the governments of Mexico, Canada, and the United States must "work together to resolve issues and ensure responsible use of scarce resources and the free flow of both resources and capital across all three borders."

Building a North American Community served as the founding document for the highly contentious Security and Prosperity Partnership of North America (SPP). The SPP, a sort of NAFTA-plus, commits the governments of Canada, Mexico, and the United States to European Union–like cooperation in what SPP documents describe as "markets and democracy, freedom and trade and mutual prosperity and security." One of the SPP's first acts was to establish an Oil Sands Expert Group to study pipeline and market issues "associated with the value added development in Canada of the oil sands." The U.S. government website for the SPP formally declares "greater economic production from the oil sands" as a central goal of energy integration. At a 2006 gathering in Houston, Texas, the Oil Sands Expert Group agreed to examine "options and plan for a smooth transition towards bitumen production that could be as high as 5 million barrels per day." In the spirit of integration, a contingent of Mexican energy employees attended the meeting too.

A 2006 report by the U.S. Congress saluted the tars sands as a "new force in the world oil market." It explained that "the proximity of this growing source of supply is a highly positive development for the U.S. and indeed the world." To replace Persian Gulf imports alone, the United States would have to drain all of Canada's projected crude production by 2016: 3.8 million barrels a day. "North American energy independence thus would require a dramatic ramp-up in oilsands production far beyond any of the current projections," the report concluded.

U.S. Energy Secretary Samuel Bodman declared in 2006 that "the hour of the Oil Sands has come" and that much integration would follow. He explained that twenty-two pipelines, thirty-four natural gas pipelines, and ninety-one electric transmission lines already linked the northern mouse with the southern elephant and said those numbers were sure to increase.

The rapid development of the Alberta tar sands has also served as a dirty-oil laboratory. Utah has 60 billion barrels of tar sands that are deeper and thinner, and therefore uglier, than Alberta's resource. To date, appalling costs and extreme water issues have kept Americans from ripping up 2.4 million acres of western landscape. But that may soon change. "Those who doubt that unconventional fuels are economically viable probably are suffering from a neck ailment that keeps them from looking north," observed Republican Utah Senator Orrin G. Hatch in 2006. "The 800 pound gorilla is sitting just above Montana, and let's face it, it's hard to miss... It's a gigantic success story and it began with Alberta's government deciding to promote the development of this resource and not giving up." U.S. companies active in the tar sands, said Hatch, "are only waiting for the U.S. government to adopt a policy similar to Alberta's which promotes rather than bars the development of the unconventional resources."

Hatch wasn't the only one to notice the Alberta gorilla. In 2006, a three-volume report by the Strategic Unconventional Fuels Task Force to the U.S. Congress gushed that Alberta's rapid development approach to "stimulate private investment, streamline permitting processes and accelerate sustainable development of the resource" was one that should be "adapted to stimulate domestic oil sands." Even with debased fiscal and environmental rules, though, the U.S. National Energy Technology Laboratory has calculated that it would take thirteen years and a massively expensive crash program to coax 2.4 million barrels a day out of the U.S. tar sands. A 2008 report by the U.S. Congressional Research Service candidly concluded that letting Canada do all the dirty work in the tar sands made more sense than destroying watersheds in the U.S. Southwest: "In light of the environmental and social problems associated with oil sands development, e.g., water requirements, toxic tailings, carbon dioxide emissions, and skilled labor shortages, and given the fact that Canada has 175 billion barrels of reserves...the smaller U.S. oil sands base may not be a very attractive investment in the near-term."

The hard work of U.S. policymakers and security experts has often been dwarfed by Alberta's consummate salesmen. Alberta has marketed rapid tar sands development with more gusto than a Bay Street broker. In several visits to Washington, D.C., as premier, Ralph Klein posed for photos by monster trucks and declared that the province had "energy to burn." He also said that he'd never met a pipeline he didn't like. Whenever critics such as Al Gore raised concerns such as the tar sands' carbon making or water guzzling, Klein would make a similar reply: "The United States needs our oil. I don't know what he proposes to run [the country] on, maybe hot air?"

Klein sent his former energy minister Murray Smith, also a dedicated oil man, to Washington to preach the tar gospel. In 2006, Smith gave a remarkable speech to members of the Interstate Oil and Gas Compact Commission in Austin, Texas. He said that the tar sands were part of a northern energy corridor, which included Alaska and the Northwest Territories, that "was going to supply energy to this continent for the next 100 years." He vowed that tar sands production would not only replace depleted U.S. oil stocks (a shortfall of some 340,000 barrels a day) but "fuel economic growth." Powering the continent, he implied, was as simple as taking two tons of sand, adding hot water, and mixing briskly. He added that an open-pit tar mine moved enough earth every day "to fill Yankee Stadium...and some would say that's not a bad idea."

Compared to a conventional oil well, Smith said, which peters out in a couple of years, the tar sands are almost inexhaustible: "You mine it, you build it, cash flow it for 30 years and then it drops off and finishes." The production of five million barrels a day was no pipe dream, according to Smith. He boasted that the province had issued 2,700 tar sand leases, and there was lots more bitumen left. The "royalty structure for oil sands is [that] we 'give it away' at 1 per cent," he emphasized. Companies don't pay 25 per cent on the price of bitumen until they've paid off their entire capital costs.

Smith ended his talk with a direct appeal for mobile workers. Although thousands of foreign workers already toiled in the sands, Alberta needed more labour: "If any of you have children who are engineers, I would like you to entice them to work in the naturally air conditioned comfort of Fort McMurray as opposed to this oppressive humid environment of Austin." The resource belongs to Alberta, concluded Smith, "but the opportunity belongs to all of North America."

What Smith advertised as North America's opportunity has arguably become a provincial debacle and a national fiasco. Like a bungled bank job, the rapid development of the tar sands has careened into a string of morally questionable decisions that could well undo the country, if not the continent. Social critic Wendell Berry once observed that "there are such things as economic weapons of massive destruction," and the rapid development of the tar sands is one of them. Dangerously, it appears to be a hydrocarbon invasion still gathering force. Given that only 3 per cent of the accessible bitumen has been recovered since 1970, most tar sands analysts contend that "the oil sands industry is just getting started."

FOUR

HIGHWAY TO HELL

....

"Frontier expansion without adequate planning has left cities
crippled by shameful environments which cause human casualties."

ELDEAN V. KOHRS, PSYCHOLOGIST, SPEECH TO

THE ROCKY MOUNTAIN AMERICAN ASSOCIATION OF

THE ADVANCEMENT OF SCIENCE MEETING, 1974

THE HIGHWAY TO Canada's El Dorado formally begins about 125 miles
north of Edmonton, just past the busy Al-Pac pulp mill and a village
called Amber Valley, where Oklahoma blacks once settled to make a
new start. The government originally built the road to serve Fort
McMurray, then a mining community of 25,000, in the 1970s. Now the
150-mile, all-weather Alberta autobahn accommodates a population of
more than eighty thousand fortune seekers on the move. Every day,
nearly fifty newcomers travel north on the highway, which snakes
through spruce and muskeg to the mines. They don't know that most
people in Fort McMurray call the road Hell's Highway, Suicide 63, or
the Highway of Death. The police call it McMurray 500.

Highway 63 is not as dangerous as the North Yungas Road in Bolivia,
an Andean precipice that sends hundreds of motorists to their deaths
every year. Nor is it as unreliable as the Siberian road to the oil fields of

Yakutsk, which becomes a deep bog every spring. But Hell's Highway offers its own set of challenges. Even before the current boom, miners travelled the long road warily, in heavy vehicles with bumper stickers that read, "Pray for Me. I Drive Highway 63."

Since 1996 and the *Declaration of Opportunity*, traffic on the road has increased to a frantic level, as has the praying. In 1999, the highway killed three workers. By 2001, the number had shot up to nine. In 2007, Highway 63 claimed seventeen lives. Between 2001 and 2005, one thousand collisions killed twenty-five people and injured nearly three hundred more. Every week, regular as clockwork, the highway silences or maims another miner.

On any given day, thousands of logging trucks, SUVS, semitrailers, buses, and tanker trucks form a nonstop parade to and from the mighty tar sands. Convoys carrying extra-wide loads, including tires and coker ovens the size of houses, often take up three-quarters of the highway. These megaconvoys move at ten miles an hour and effectively block any view of oncoming traffic. According to Syncrude Canada, Highway 63 probably ferries the highest tonnage per mile of any road in Canada and is "inadequate for the traffic that uses it." TransAlta once dumped a steam turbine on the highway.

The road's inadequacy encourages a certain do-or-die recklessness. Drivers pass not only on solid lines on hills but also on soft shoulders, at speeds that might alarm racecar professionals. (The average speeding ticket clocks in at nearly a hundred miles an hour.) Impatient drivers regularly swing onto the shoulder to catch a glimpse around a wide load, then dart out into the other lane to pass like bats out of hell. You never know when your number might come up.

Thursday and Sunday evenings are the worst. That's when the shifts change at the mines and thousands of workers return to their families and girlfriends in Edmonton. Most are exhausted; many are drugged on amphetamines or pissed to the gills. A lot of people won't drive at all on those days, particularly with children. They don't want to be remembered as another little white cross decorated with a blue hard hat, an empty Russian vodka bottle, or an overstuffed teddy bear along the roadside. Every week the local newspaper, *Fort McMurray Today*, reports another bloody accident due to "burgeoning oilsands development."

Even Alberta politicians, who celebrate the energy boom as if it were a grand birthday party, openly fear Highway 63. When former transport minister Lyle Oberg, a physician, went to check out the traffic a few years ago, he spent most of his visit patching up an accident victim. Whenever local MLA Guy Boutilier drives "the zoo," his wife, Gail, anxiously awaits his return, greeting him at the door with "Thank God you made it." After a bus accident killed six workers and injured eight in 2005, the government championed a $650-million plan to twin the road. Three years later, the twinning has grown into a $970-million project. Given labour shortages and other inflationary pressures (the road will cost two-and-a-half times more per mile than any other road in Alberta, because of the muskeg), no one expects it to be finished any time soon.

The carnage on the road, like everything about the tar sands, is graphic. It's heavy metal on heavy metal, and at high speeds. When a worker struck a wide load near Mariana Lakes in 2007, it sheared off the top of his car. After a crash involving two semitrailers in 2008, firefighters spent a couple of hours removing the mangled bodies from a cab that looked like an accordion. In one notorious accident, a logging truck clipped the back of a parked flatbed trailer. The collision pitched the truck's logs missilelike into an oncoming minivan carrying sixty-two-year-old Ralph Brandson and thirty-seven-year-old Erkin Kanhodjaev, an immigrant from Kazakhstan. The logs crushed both men. In recent years, the International Brotherhood of Electrical Workers has lost thirteen members on Hell's Highway. The RCMP issued nearly eighteen thousand driving violations in 2004 on one stretch of the highway alone.

Officially, Alberta Energy spokespeople blame moronic drivers and wildlife for the mounting death toll. Deer and moose *have* been known to take suicidal runs at semitrailers and Ford F-350s on Highway 63. But most McMurrayites quietly concede, as one Internet blogger wrote, that "people on Highway 63 drive like assholes." Most agree "there should be a HUGE premium paid by the large oil companies for all of the HUGE loads being hauled up that highway and destroying it." Petropoliticians, however, never talk this way.

Muriel McKay speaks plainly, as most northerners do. McKay and her husband, Steve, run Mariana Lakes Country Store, an hour's drive south of Fort McMurray on a stretch of road where wildfires have turned swampy spruce trees into a mess of black toothpicks. The busy store used to be a twenty-four-hour restaurant, gas bar, and eighty-man camp for the natural gas drilling business. Alarmed by tar sands workers who would attack Hell's Highway after too many drinks, McKay closed the bar in 2000. "We didn't want to contribute to drinking and driving. It was a moral decision," she says. Shortly afterwards, the couple closed an off-sales liquor store that had sold $200,000 worth of booze a year. Too many workers were buying beer or bourbon after shift changes and then hitting the road. No one knows how many lives the McKays have saved, but the number is substantial.

McKay, who grew up in this tamarack-and-spruce country, says she's never seen an economic boom like the current one: "It's mindboggling and overwhelming. I can't get my head around the figures." Nor does she believe that the traffic, the construction, and the forest-eating turmoil will slow down, since they are driven by the global addiction to oil. "Are you going to stand there and try to stop a tsunami?" she asks. "Can you stop a tidal wave?" In the last two years, there's been endless talk like that in Fort McMurray. Just about everyone in town believes that the rail link between Edmonton and Fort McMurray should have been upgraded years ago and that industry should have paid the bill.

FORT MCMURRAY, THE tar sands capital CEOs call "the anchor of prosperity," sits near the middle of the Regional Municipality of Wood Buffalo, at the confluence of two boreal rivers: the Athabasca and the Clearwater. The municipality, which encompasses a 26,000-square-mile forest the size of Tasmania, is both a wilderness and North America's busiest industrial centre. Fort McMurray used to be surrounded by trees, but tar sand leases will soon surround the city's neighbourhoods. The Chamber of Commerce predicts that its city will be the largest community in the world north of the fifty-fifth parallel by 2100 because every job in the mines generates three service opportunities.

White people once described Fort McMurray as a "fur factory." Now bureaucrats call it an "island of developable land surrounded by muskeg." The average house costs more than $600,000, the highest price in Canada, so it's a good thing the average annual income hovers around $100,000. People who make less than $70,000 up here, about 30 per cent of the population, live below the poverty line. The average age of a bitumen fortune seeker is thirty-one. Says Cheris, a petite twenty-nine-year-old pilot from Saskatchewan, "I've never seen such a transient population in all of my life. We are mostly young, make lots of money, play hard, and then go home. We aren't contributing anything."

Newcomers and visitors generally stop at the tourist office just off Highway 63. The cheerful women operating the place bubble with information. Lisa Ashley has raised two children in the city and witnessed a doubling of the population in a decade. She says the most common question at the visitors' booth is "Where do I apply for a job?" Every day the office gets calls from as far away as Germany, Brazil, and Norway. The Norwegians want to know about farming opportunities in the tar sands.

Entries in the visitors' book capture the frenzied character of the boom. One couple complained they couldn't find a campground in the summer because transient workers occupied them all, but Wal-Mart, "thank goodness," let them camp in the store parking lot. Most visitors found that the open-pit mines "exceeded our imagination" or were "totally awe inspiring." A fellow from the oil shale region of Colorado jotted, "Wow. Maybe seeing the future here." One retired couple from Ontario wrote, "We have been to Dawson City, an ancient boomtown. Now we want to see a boom town actually booming." A Japanese pair who came to conceive a child under the northern lights, for good luck, just complained about the cold.

Fort McMurray has an untidy yet familiar global face. The people who run the mines are generally either confident Calgarians or engineers and managers with Oklahoma and Texas accents. Most of the multinationals employ professionals from Venezuela, India, China, England, and the Middle East. Nannies from the Philippines take care of the kids while taxi cab drivers from Ethiopia or Somalia transport the transients. The Ethiopians describe the winter cold as a "bullet" and dream of starting businesses elsewhere. Muslims, who built one of the world's most northerly mosques in Fort McMurray, run many of the camp kitchens. About half of the general workforce hails from Newfoundland or the Maritimes, the poorest parts of Canada. The rest come from struggling rural communities throughout North America. Many companies fly in hundreds of temporary workers from China, Mexico, and Croatia, too, and when the mines kill a temporary man or two, investigations can take a year to complete. The homeless in Fort McMurray are generally Cree or Dene. They once hunted moose and trapped beaver where the mines and upgraders now stand.

Ruth Kleinbub and Grant Henry moved to Fort McMurray from Ontario nearly thirty years ago. They raised four children on the banks of the majestic Clearwater River in an older part of town called Waterways and went to the dump every Saturday for entertainment. Their neighbour, a Métis trapper, taught the kids how to hunt rabbit and weasel. The odd wolf loped by, as did a mother bear. "You'd always see wildlife on your way to work," recalls Grant, a welder.

But the bitumen boom has erased that Fort McMurray. The city now shoots wandering black bears. No one parks their dog teams out front of the IGA because they would get run over. Finding a familiar face downtown is no longer a given. "McMurray was a northern community, and now it's a city with southern ambitions," says Grant. The fishing has gone to hell and the rush-hour traffic is "overwhelming." Nearly five hundred homeless folks shuffle about the windswept streets downtown. "The ways they grew up with have been destroyed," says Grant. Ruth, a well-known local environmentalist (a rare species here), offers a short list of concerns about the boom: "The land, water, and air."

Writer Wallace Stegner recognized that there was nothing as exuberant or as adolescent as a boomtown. He reckoned that the West's resources and scenery attracted two kinds of people: stickers and boomers. The stickers felt an affinity for place and believed in giving as opposed to getting. Many of the people who settled in Fort McMurray in the 1970s and 1980s became stickers and community makers. But the rapid development of the tar sands since 1996 has brought in wave after wave of boomers. The tension between the two tribes is palpable. "They like to make the money, but don't have anything good to say," observes one businessman. None of the stickers thinks the bitumen boom has improved the city's character.

Boomers aren't interested in making a living; they want to make a killing. As Stegner noted, they behave in the frontier like children in a candy store. With a shadow population of twenty thousand camp workers as well as itinerant engineers filling the hotels, boomer culture is transcendent in Fort McMoney. You fly in, work, and fly out. Even environmental critics zip in and out like black flies. Everyone expects it to be that way for decades to come. Ninety-eight per cent of the population says they will retire somewhere else.

The Cree, who have survived fur, gold, and uranium booms in the region, have an interesting name for shiftless white folks: *Namoya Nehiyaw*. Because the Cree still know their way around a boreal forest, they call themselves *Nehiyaw*, or "a smart person." *Namoya* means "no" or "not." So a *Namoya Nehiya* is a not-so-smart person.

Just about everyone you meet in Fort McMurray talks about the money. A Dene woman who cleans rooms at a work camp calls the place "a twenty-four-hour shift town" where people make money and "take it all home, east or west." A lonely Lebanese woman from the Bekka Valley with three children says, "There is lots of money in this place and that's it. There is no village." A nineteen-year-old girl who makes $27 an hour at one of the mines comments that "the most extreme thing is the money and how it's taken over people."

Fort McMurray hums like San Francisco during the gold rush, bustles like Los Angeles during the oil boom of the 1920s, hustles like

Dawson City during the Klondike rush. In 1974, U.S. psychologist ElDean V. Kohrs coined the term Gillette Syndrome to describe the social wreckage caused by a coal boom in Gillette, a bucolic Wyoming ranching community in Carbon County. Kohrs wrote that a "history of power production— synonymous with boom development" usually left behind "a dismal record of human ecosystem wastage," including "spiritual depression, divorce, drunkenness, dissension and death."

Kohrs pegged the dark side of a resource boom unerringly. Divorce rates went up because "fatigued men working long shifts and driving long distances to work came home to equally fatigued wives coping with a mud spattered world." Young people left school early to get high-paying jobs. Transients who worked hard and played harder filled the jails. Medical care went to hell, because people found it difficult to get a family doctor and resorted to the hospital emergency room for routine treatment. Trailer parks sprung up with no regard for people's "psychological well-being." A frontier carbon society, said Kohrs, "has no use for those who are not productive in their narrow definition and this implies full vigor and health."

Kohrs concluded that hydrocarbon hurricanes "wreaked a toll of human suffering, developed communities that flared in the boom, blazed and died, or flared and continued wildfire growth without care or planning, leaving wakes as devoid of *quality* of life support as a prairie grass burn area." Community leaders, he warned, must "support psychological well being for all, not merely a few who can escape the human crises of boom because of their wealth."

Thanks to peak oil and the demand for natural gas, many communities in the west, from Fort McMurray to Farmington, New Mexico, are suffering a bad case of the Gillette Syndrome. Alberta's divorce and school dropout rates are among the highest in the nation. According to Statistics Canada, Alberta women experience the highest level of spousal abuse in Canada. Alberta's premier, Ed Stelmach, calls such inconvenient facts "the price of prosperity."

In 2006, the Athabasca Regional Issues Working Group, a nonprofit industry association, released a report on "sustainable community indicators" in Fort McMurray. Not surprisingly, it highlighted what Kohr would call "less civilized living conditions." The group reviewed twenty-one quality-of-life indicators, including affordable housing, crime, voter participation, residential water use, and traffic collision rates. Nearly half were "worse" or "worsening." In 2008, the city cancelled its Blueberry Festival for lack of volunteers.

Criminal Intelligence Service Alberta, a government agency that shares intelligence with police forces, reported in 2004 that the boom had created fantastic opportunities for the Hell's Angels, the Indian Posse, and other entrepreneurial drug dealers: "With a young vibrant citizen base and net incomes almost double the national average, Fort McMurray represents a tremendous market for illegal substances." By some estimates, as much as $7 million worth of cocaine now travels up Highway 63 every week on transport trucks. According to the *Economist*, a journal devoted to studying global growth, about "40 per cent of the [tar sands] workers test positive for cocaine or marijuana in job screening and post accident tests." Health food stores can't keep enough urine cleanse products in stock for workers worried about random drug trials. There is even a black market in clean urine.

Fort McMurray reports five times more drug offences than the rest of Alberta, because ordering crack cocaine at a work camp is easier than ordering a pizza. The boomtown also has an 89 per cent higher rate of assault and a 117 per cent higher rate of impaired driving. Since 2006, the local office of Alberta Alcohol and Drug Abuse has witnessed a 25 per cent increase in the number of clients. According to a recent provincial report on rapid tar sands growth, "some employers are more tolerant of alcohol and drug abuse simply because there is no one else available to do the job." Some contractors swear they'd lose half their crews if they did drug testing.

Housing in Fort McMurray is scarce and expensive. It's a major example of freak economics. The province owns all the land in the city, but it has managed sales poorly. As a result, the price of a single-family home has climbed from $174,000 to more than $600,000 in a decade.

That's twice the average price of a home in Canada. Even a mobile trailer can cost $300,000.

In 2005, Alberta's auditor general, Fred Dunn, investigated the way the government sold land in Fort McMurray and exposed a Third World mess. Dunn found that the Crown corporation charged with selling land didn't have a plan. It timed sales sporadically, could not show that it received fair value for four of nine sales audited, relied on appraisals without verifying them, and "did not use an open and transparent process for three of its land sales." In one case the corporation gave land away free to a developer. Dunn didn't have a mandate to take his investigation further.

Such negligence and outright corruption have also helped to drive rents sky high. A bachelor suite in Fort McMurray goes for a spectacular $1,300, and as many as nine people might share an apartment to keep the costs down. Some workers pay $700 just for a cot in a walk-in closet, while others rent sheds or garages. There is no privacy.

In summer, homeless Aboriginals and crackheads sleep under cars or in tents amid piles of garbage by the Syne, a small channel of water near downtown. One heavy equipment operator admitted to sleeping in his truck to save money. Almost every house in the city has four or five cars parked out front; relatives and friends often stay longer than anyone expected. In the wintertime, many tradespeople will plaster a Cardinal, Colorado, or Titanium trailer with bubble wrap, tarps, and insulation, then camp out in −40°F weather and pay $30 a day at a trailer park for the privilege.

The housing emergency has even become a crisis for cats and dogs. Every day, newcomers and families looking for work arrive in town not knowing they won't be able to afford a house with a yard or find an apartment that allows pets. The local SPCA, which can shelter a hundred animals, now houses two hundred and has run out of room. Last year, staff turnover was 300 per cent. Just about every social agency reports similar crowding and staffing issues.

Fort McMurray's population has multiplied like a cancer cell, with annual growth rates of between 9 per cent and 12 per cent. In 1999,

36,000 bitumen diggers, upgraders, and assorted helpers lived in the mining town. Eight years later, more than 65,000 crowded the place. During the same period, the population of the Regional Municipality of Wood Buffalo shot from 42,800 to 90,000. An immigrant from Yellowknife says she doesn't know where people are going to live, go to school, or shop when the population hits the projected 100,000 in 2011. Nobody does, really. CNRL and Shell have built their own airstrips, the largest private runways in Canada, as well as decent camp housing. They fly their employees in and out every two weeks. To avoid the congestion and mess of Fort McMurray, Imperial Oil has talked about building another city forty-five miles north of town.

Fort McMurray's growth rates are "exponential." The father of peak oil, Marion King Hubbert, once noted that exponential growth is characterized by doubling and that just a few doublings can lead to hellish numbers. That's a reality now evident to everyone in town. A 3 per cent rate of economic growth will lead to doubling in twenty-three years. A 10 per cent rate, such as Fort McMurray's, doubles in just seven. Each successive doubling consumes as much energy and resources as all the previous doubling periods have done. Exponential growth explains both why the world is running out of cheap oil and why Fort McMurray has become an urban nightmare with an infrastructure deficit of nearly $2 billion.

Inflation is rampant here. Most call it the Fort McMurray Factor. When city council undertakes any public infrastructure project, it ruefully adds 45 per cent to 55 per cent to its cost. A beautiful recreation complex budgeted at $23.4 million in 2005 ballooned to more than $200 million, and the city doesn't know if it can afford to operate the facility. A modest police headquarters started at $10 million but grew to $50 million. The regional landfill, which started at $13 million, became a $24-million endeavour. The water treatment plant went from $39 million to a staggering $218 million. Not surprisingly, the Regional Municipality of Wood Buffalo has the highest debt load of any municipality in the province. It soon won't be able to borrow any more money. Outrageous cost overruns for industry are also the norm.

Service, meanwhile, has become a dead custom in town. It takes forty minutes to order a cup of coffee at Tim Hortons. Lineups at the banks on payday can be sixty people deep. McMurrayites lament daily that they must wait for everything. Almost every store has big signs announcing "zero tolerance" for abusive customers, which is comical given that hardly anyone actually bothers with service. The Oil Can, the kind of blue-collar saloon that enlivens any mining town, warns visitors that it won't tolerate "fighting or rowdy behaviour," "physical assaults on staff," theft, or drug dealing.

Young couples such as Darrell and Heidi regard Fort McMoney as the North American dream on overdrive. Heidi, whose father worked at Syncrude, grew up in town, and Darrell, an Edmontonian, has lived there since the age of nineteen. The gregarious pair live in a seventy-two-foot Ridgewood manufactured home with their two young children in Prairie Creek, south of town. Heidi used to work at the Safeway but now stays at home raising the kids. She can't believe the frantic nature of the city and says that you might wait nine hours for medical attention at the hospital's emergency department. Almost everybody in town knows somebody who has had cancer, has died of cancer or committed suicide. "If our marriage can survive McMurray," she says, "it can survive anything."

Darrell, a small businessman and entrepreneur, believes you can't really understand Fort McMurray unless you live there. Things are so fast-paced that "if you snooze, you lose." The boom has so inflated wages that it has gutted the work ethic and destroyed many small businesses, he says. "Employers are grateful if a guy shows up. They say, 'I've got a good guy. He showed up when he said he would.' That makes me laugh." Many men also spend obscene amounts of money on fossil fuel–burning toys. Every Ralph and Ed in town, Darrell says, seems to "own 2.2 cars, 2.2 quads and 2.2 snowmobiles."

Darrell bought his Ridgewood home in 2000 for $140,000. He planted it on a pie-shaped lot about 115 feet in length. Today the trailer, a roomy affair, is worth more than $400,000. Darrell thinks "there is more money than brains" in the city, but he admires the spunk and resourcefulness of its families and its businesses. "Fort McMurray has some of the best people I've met in a long time. This place is not fuelled by the mines but by the people."

Yet Darrell reckons 99 per cent of McMurrayites don't have a clue how big the resource is or how big the boom could get. "The environment is 495 on a list of 490. It's not there. Most don't care. It's not home. It's a place to hang your hat and make money, and you go with the flow or get run over."

A young man in Fort McMurray can experience life at its best and its worst. In 1999, Ted, then a thirty-one-year-old heavy equipment mechanic, moved up here to pay off his credit card debts. He stayed for six years and swears he has never seen a place as crazy, as filthy, or as corrupt.

Ted laboured in the mines as well as for heavy equipment contractors, working for twenty-four days straight on ten-hour shifts before getting four days off. One employer offered him $50,000 a month if he would deal cocaine at work camps; he refused. The city's drug problems are so bad, he says, that if "they did a mandatory check on everyone who worked in the mines, they'd have to close the place down." The first question ambulance crews ask men with chest pains is "Are you on cocaine?" Ted married a woman from Newfoundland but divorced her after she got hooked on drugs and went wild partying. "The human mind can't handle rags to riches. It can't go from welfare to a $150,000 job and not go sideways," he says.

The social hierarchy in town is fixed and elaborate. Managers and engineers from Suncor and Syncrude get the best spots in the clubs and sports arenas. Next in line come Shell's people. "Everybody hates CNRL [Canadian Natural Resources Ltd.]," says Ted, because they stole workers from the other firms. Nobody talks to the foreign workers from the depleted oil fields of China, "because they behave and work like railway coolies." Construction workers lie at the bottom of the heap. Everyone, says Ted, is paranoid and angry about something.

In 2002, Ted bought a house in Timberlea, where pollution corroded his brass door fixtures and left a film of soot on his windows.

Until he left Fort McMurray, Ted says he had constant string of sinus infections, pink eye, breathing problems, and a dozen strep throat infections. "Everyone had it. But greed is more important than your own personal health."

Almost half of the city's population comes from the failed fishing communities of Newfoundland. Fresh-faced young women from the Rock can be found waitressing at most restaurants, where they confide to customers, "I want to go home." For thirty years now, Newfoundlanders have dug bitumen for the U.S. export market. They jokingly call Fort McMurray Newfoundland's second-largest capital and eat at the Kozy Corner, which serves cod tongues and tasty fish and chips.

When Sue Pearce, a Newfoundland émigré and union representative, moved to the boomtown three years ago, three things surprised her. "The first thing was the twelve-hour shifts. That was a surprise, to have people away from their home such long periods of time." The second was the Filipino nannies who look after the kids because the parents are working those long shifts. "That was a shock." And the third was the scale of the industry in the bush. "Oh, my Lord. It looks devastating and smells so bad. But it's the smell of money and what we do, and the companies promise it will be fixed up."

Pearce, who lives with her family in Thickwood Heights, where foxes still play near the driveway, explains how a people once dependent on cod, a renewable resource that government policies destroyed, now have a destiny with bitumen, a finite resource that the mining companies will liquidate in forty years. "We have always been survivors. We persevere. Whatever it takes to make a living, we are going to do it. We have left home for hundreds of years. Newfoundlanders built the World Trade Center. If you look at the history of steel work, that's our history too."

Many families live in neighbourhoods that resemble Calgary suburbs on streets named Wolverine and Wapiti. But camp workers sleep in eight-by-twelve-foot rooms in trailer camps in the bush, north or south of town. Some camps have all the charm of penitentiaries while others call themselves lodges. Workers say the newer camps are cleaner and more comfortable and even boast their own bars and recreational centres. The older camps have mould on the walls and shared toilets for forty-five people.

Staying in a camp for months on end will turn a guy numb. For security reasons, the camps don't allow workers to have barbecues or bonfires. The guards, generally not the type to run marathons, have the final say on who comes and goes. Camp chefs and cooks often deal drugs on the side. Everyone knows where the crack cocaine comes from, but as one camp worker put it: "You're more likely to get arrested for speeding in Fort McMurray." Industry surveys say that 75 per cent of the camp workers would never consider moving to Bitumen City and think the only thing that could improve life would be more flights back home.

Mike and Ken, two fifty-something Albertans, live in bush camps while they work on construction and pipeline jobs. They call the world's largest energy project a "shit hole" and shamelessly admit, as everyone does, that they are here for the money. The two men explode with tales, about a co-worker fired for having blisters on his lips from smoking crack and a drug dealer stabbed to death by a camp cook. Ken says he worked beside a fellow who climbed to the top of a three-hundred-foot-high coker at Suncor one night and dove off, as free as a bird, out of despair or loneliness or just plain madness. The next day the company reported that the man had "died of natural causes," even though he was flatter than a pancake. "The companies never tell us anything," says Ken. "The greed really bothers me."

The two Albertans work mostly with Maritimers, who make up nearly half the itinerant camp force. There is even a Local 420 from Cape Breton. The men work for 420 hours, then quit to go back home and collect welfare.

Liz Moore, an eighty-five-year-old Colorado grandmother who visited in 2006 to learn more about America's number-one oil source, went home appalled. Moore, who used to work at the Appalachian Regional Commission, knows a lot about open-pit mines. "This is exactly what I saw in Appalachia," she says. "It was godawful there and it's godawful here." She couldn't believe that the local museum, the Oil Sands

Discovery Centre, would raffle off Hummers or that parked trucks would outnumber cars on the street five to one. Moore took lots of pictures and set up a Web site documenting her stay. Her website, www .oilsandsofcanada.com, announces that Syncrude creates the same amount of greenhouse gases as all the coal-fired power plants providing electricity to Chicago, and that Syncrude uses enough water from the Athabasca River to satisfy a third of Denver, a city of half a million people.

Syncrude's legal department and the Alberta government didn't like Moore's online travelogue. They demanded that she remove her photographs, because she had taken them while on a Syncrude tour and had signed a form saying the images belonged to the company. Moore reluctantly obliged, then put up better photos. "We are attached at the navel with what's going on in Canada," she says, "and I don't think most Americans know it."

ONE OF THE first people to document the "human ecosystem wastage" in Fort McMurray was Dr. Michel Sauvé, a lean and fiercely articulate internist. As the president of the Fort McMurray Medical Staff Association, Sauvé appeared before public hearings run by Alberta's Energy Resources Conservation Board in 2003 and 2004. CNRL and True North Energy wanted approvals for another $12 billion worth of mining projects, but Sauvé didn't think the existing hospital and staff could handle more itinerant workers.

Industry tried to silence Sauvé. True North Energy offered the association $100,000 for health research, but only if Sauvé promised to stop talking about how more camp workers would strain the health care system or how 30 per cent of the community lived below the poverty line. The association refused, and Sauvé presented the board with facts it didn't want to hear.

After eight years of rapid tar sands approvals, Sauvé reported, Northern Lights Health Region had fewer hospital beds than a third of its provincial counterparts. Thirty-eight physicians served sixty thousand people, a physician-patient ratio of 1 to 1,579—three times lower than that of Argentina, China, Mexico, Mongolia, and Uzbekistan. The suicide and fatal self-injury rates had risen between 31 per cent and 101 per cent above provincial averages. The residents of Wood Buffalo also had "among the shortest life expectancies in the province." Why did the region with the highest social problems have the lowest per capita funding? asked Sauvé.

The ERCB ignored the statistics and approved the projects anyway. The regulator's persistent disregard for the health consequences of its approvals convinced Sauvé that the whole system was corrupt: "Consultants for corporations said everything was fine. It was disingenuous and a downright deceitful presentation of the facts. The ERCB bases its decisions on information that is entirely one-sided."

After the hearing, True North Energy refused to pay the legal costs of interveners, as required by the ERCB. Instead, the largely U.S.-funded developer used its deep pockets to fund an appeal to the Alberta Court of Appeal, arguing that Sauvé had no right to speak on behalf of local citizens because his medical association wasn't registered under the Societies Act and therefore was not "directly affected" by the project. The legal battle, which threatened to bankrupt Sauvé, lasted three years and was eventually settled out of court in the physician's favour. Sauvé donated the money to environmental causes.

By 2006, the condition of the region's health care system was even more critical. More than a third of people had to seek medical help outside the region, because Wood Buffalo now had the highest number of patients per physician in the country: 4,500. (The World Health Organization recommends 600.) By then the region had fewer doctors and nurses than in 1994, and a third fewer than the northern territory of Nunavut. In an angry letter to the ERCB, Sauvé wrote, "We have the dubious honour of having now achieved ghetto status for healthcare access ironically in the heart of Alberta's economic engine."

In July 2007, Sauvé provided an update in a letter to the *Canadian Medical Association Journal*. Anywhere between 30 per cent and 40 per cent of the population of Wood Buffalo still did not have a family doctor, he reported. Every year, nearly half of the region's hospital and public health staff left, because they either couldn't afford housing or couldn't stand the chaos. There was a 20 per cent vacancy rate for nurses. Only two of the fourteen family doctors serving eighty thousand people still accepted patients. At the local emergency room, one doctor might see a record 156 patients during a twelve-hour shift. "In the minds of many, being a boomtown translates into a run on gold faucets and line-ups at luxury car dealerships. Nothing could be further from the truth as Fort McMurray is discovering," wrote Sauvé.

In 2007, the province paid out-of-town doctors to make day trips to the tar sands city for $1,200 plus fees. In 2008, Alberta doubled those wages to entice out-of-towners to work under difficult, often dangerous conditions. One guest doctor worked fourteen consecutive twenty-four-hour shifts at the Fort McMurray hospital because there was no one to replace her. During this unprecedented medical marathon, the doctor supervised up to fifty-three sick hospital patients at one time. The national standard recommends fifteen sick patients per hospital doctor. Every day doctors and nurses experience physical or verbal threats from transient workers often high on drugs.

Although the region generates approximately $5 billion in revenue for the federal government every year, little federal money has been reinvested in the community. In October 2006, Melissa Blake, the city's spunky mayor, went on the offensive, bluntly defining the scale of social and political abuse that the boom had visited upon her community for the House of Commons Standing Committee on Finance: "Our wastewater treatment needs exceed capacity. Our water treatment plant will be at capacity next year. Our recreational facilities are over-taxed. Our landfill site is full. Fort McMurray is 2,800 housing units short of current demand. Our health care system needs a 100 per cent increase in onsite doctors."

Shortly afterwards, the Regional Municipality of Wood Buffalo intervened in three separate public hearings on multibillion-dollar bitumen projects proposed by Suncor, Shell, and Imperial Oil. Law-yers for the municipality noted that "continued approvals of oil sands projects, without addressing quality of life issues, have in the RMWB's view, tipped the scale away from the public interest." The submissions cited a long list of public concerns, including "deforestation and loss of related jobs; water quality and supply; quality of life; loss of traditional practices and places; cost of living/access to affordable housing; excessive change; access to and adverse impacts on services including education, policing and health care." The lawyers argued that though the regulator had acknowledged the problems repeatedly since 1999, neither industry nor the provincial government had done anything.

The ERCB listened politely and then approved all three projects, one after the other. It ruled that the risks to air, water, and human health were "acceptable" and that everyone should "adaptively respond" to the region's corporate anarchy. In a rare sober moment, the board admitted that "the capacity of existing infrastructure, which in effect has facilitated the expansion of the oil sands industry to this point, has been depleted." It also found an "apparent lack of a coordinated response among government departments and various levels of government." In other words, there was no planning.

Although the mandate of the ERCB makes it responsible for "orderly, efficient and economic development in the public interest," the board has yet to explain to the RMWB, Mayor Blake, or the citizens of Fort McMurray how poor water quality, unaffordable housing, astronomic rents, hellish roads, and chronic labour shortages constitute "orderly" or "efficient" development. When a prominent Fort McMurray businessman told Brad McManus, acting chairman of the ERCB, in 2007, that the tars sands "were out of control," McManus replied, "We're the regulator. We can't say that."

Hydrocarbons define and shape the tenor of everyday life in the region. People often wake up to the smell of sulfur dioxide or ammonia. They drive everywhere in Power Wagons or Dodge Ram trucks (hardly anyone walks); eat at restaurants with names like Fuel; attend spirited hockey games played by the Oil Barons; get drunk at the Oil Can; gamble at the BoomTime Casino, and constantly drive past welcome signs that remind them, "We have the energy."

Anyone who spends a week in Fort McMurray can't help but think of Al Swearengen, a pimp, wife abuser, and gambler who kept a saloon and whorehouse in the town of Deadwood, South Dakota, during the gold rush in the Black Hills in the 1870s. Greed rode through the mud streets there as brazenly as a naked man on a horse. The HBO series *Deadwood* immortalized the character of Swearengen with a raw script. The boomtime philosopher called things the way he saw them, in all their crudeness and glory and stupidity: "Pain or damage don't end the world. Or despair or fucking beatings. The world ends when you're dead. Until then, you got more punishment in store. Stand it like a man... and give some back." That's straight boom talk, of course. Swearengen also recognized that "change ain't lookin' for friends. Change calls the tune we dance to."

AS A TWENTY-YEAR resident of Calgary, I have watched the "human ecosystem wastage" escalate year by year, as hundreds of fortune seekers pour into my city every week. Each day on my way to work I pass another homeless man ruined by crack cocaine or bad bitumen luck. My wife is reluctant to park the car downtown. Panhandlers dot the streets. Every other week someone tries to break into one of our two vehicles. Last year a crackhead threw a rock through a passenger window to retrieve a loonie.

Just three blocks from our house in a so-called desirable neighbourhood, a man's arm was found in a Dumpster. Police found the rest of the body somewhere else. A friend's stolen van turned up nearly two hours north of Calgary. It had been used for breaking and entering: thieves plowed the vehicle through a garage door, entered the attached house, and loaded up the van like it was Christmas.

Avarice fills the Calgary air, and most people run like hamsters on a treadmill. I used to know my neighbours, but I can't keep up now with all the investors. There are too few schools in the city, and hospitals line patients up like waiting airplanes in the halls and bathrooms.

My three sons believe that driving a BMW or a Porsche is normal, because a bitumen boom fills the streets with flashy cars. The traffic is overwhelming. My property taxes have nearly doubled, and I doubt that my children will be able to afford to make a life here. I no longer have a family physician and, like a million other Calgarians, probably never will.

Though a select few of the bitumen elite are ordering $10,000 boxes of truffles from France, they don't look satisfied yet. As Christian philosopher C.S. Lewis wrote, greed is a form of pride, and "Pride gets no pleasure out of having something, only out of having more of it than the next man."

In 2007, journalist Alexandra Fuller documented the impact of the latest hydrocarbon boom in Carbon County, Wyoming. This time around, EnCana Corporation, a Canadian company, was drilling for unconventional natural gas at a Fort McMurray pace. Fuller's *New Yorker* story could have been about the tar sands. After describing the crystal meth use among EnCana's rig workers, the wife beatings, the road accidents, the destruction of wildlife, the uprooting of families, the debasement of property rights, and the whole unmitigated frenzy, Fuller concluded: "A place in the throes of an energy boom isn't so different from a person in the throes of addiction: there's the denial that things are out of control; there's the sleeplessness and the moral carelessness, and the fact that you are doing something that you know isn't good for you but you just can't stop."

Canada, a suburb of Fort McMurray, is in the throes of an addiction.

FIVE

THE WATER BARONS

....

"There's a myth out there that oil sands production
comes at too high an environmental cost. This myth has
gained some traction here in the U.S."

ED STELMACH, ALBERTA PREMIER, ALBERTA ENTERPRISE

GROUP ENERGY FORUM, WASHINGTON, D.C., 2008

TO APPRECIATE THE world-class impact of the tar sands on the globe's
third-largest watershed, it's instructive to look first at the hardwood
forests of Appalachia. That's where the coal industry has practised an
unconventional mining technique known as mountaintop removal
since the 1980s. The industry swears that the innovation is cheaper
and safer than digging underground.

Mountaintop removal and open-pit bitumen mining are classic
forms of strip mining, with a few key differences. In mountaintop
removal, the company first scrapes off the trees and soil. Next, it blasts
up to eight hundred feet off the top of mountains as ancient as the
140-million-year-old Himalayas. (In West Virginia alone, industry goes
through three million pounds of dynamite every day.) Massive earth
movers, like those used in the tar sands, then push the rock, or "excess
spoil," into river valleys, a process industry calls "valley fill." Finally,

giant drag lines and shovels scoop out thin layers of coal. Electrical consumers as far afield as Ontario and Washington, D.C., now keep their dishwashers running and their iPods charged with coal-fired electricity powered by mountaintop removal.

In the tar sands, companies specialize in forest-top removal. First they clear-cut up to 200,000 trees, then drain all the bogs, fens, and wetlands. Unlike in Appalachia, companies don't throw the soil and rock (what the industry calls "overburden") into nearby rivers or streams. Instead, they use the stuff to construct walls for the tar ponds, the world's largest impoundments of toxic waste.

As earth-destroying economies, mountaintop removal and bitumen mining have few peers in their role as water abusers. No U.S. government agency considered the cumulative impact of the mountaintop removal on Appalachian rivers and streams until 1998, when a lawsuit by one devastated community forced the U.S. Environmental Protection Agency (EPA) to tally up the damage.

The EPA published its damning findings in a series of studies, despite massive interference along the way by the coal-friendly administration of George W. Bush. In an area encompassing most of eastern Kentucky, southern West Virginia, western Virginia, and parts of Tennessee, mountaintop removal smothered or damaged twelve hundred miles of headwater streams between 1985 and 2001. (Headwater streams bring life and energy to a forest.) The studies were blunt: "Valley fills destroy stream habitats, alter stream chemistry, impact downstream transport of organic matter and...destroy stream habitats before adequate pre mining assessment of biological communities has been conducted." The EPA predicted that mountaintop removal would soon bury another thousand miles of headwater streams.

Downstream pollution from the strip mines also contaminated rivers and streams with extreme amounts of selenium, sulfate, iron, and manganese. In addition, mountaintop removal dried up an average of one hundred water wells a day and dramatically affected groundwater quality. All in all, more than 450 mountains were destroyed during that six-year period, as well as 7 per cent (370,000 acres) of the most diverse hardwood forest in North America.

The tar sands have already created a similar footprint in the Mackenzie River Basin, which protects and makes one-fifth of Canada's fresh water. Throughout the southern half of the basin, bitumen mining destroys wetlands, drains entire watersheds, guzzles groundwater, and withdraws Olympic amounts of surface water from the Athabasca and Peace rivers. A large pulp mill industry struggles along in the wake of the oil patch, and a nascent nuclear industry threatens to become another water thief in the basin.

To date, no federal or provincial agency has done a cumulative impact study evaluating the industry's footprint on boreal wetlands and rivers. However, Environment Canada knows that a day of reckoning is coming. Briefing notes for senior officials at the department obtained by journalist Mike de Souza warned in 2006 that "the lack of a proper assessment of the cumulative effects associated with these projects could result in legal challenges of federal and provincial approvals." Two years later, a federal court judge, responding to a lawsuit launched by four environmental groups, found that the environmental assessment of Imperial's massive Kearl project was so flawed that the Department of Fisheries and Oceans temporarily withdrew the project's water permit. The federal cabinet quickly reissued the permit, but more legal water challenges seem as inevitable as rising oil prices.

Just about every damn agency in the country has expressed alarm about water use in the tar sands. The Petroleum Technology Alliance of Canada, for example, a Calgary-based nonprofit research group, declares water use and reuse to be the region's biggest issue, because "bitumen production can be much more fresh water intensive than other oil production operations." The National Energy Board, no radical group, has questioned the sustainability of water withdrawals for bitumen mining. The Department of Fisheries and Oceans says that the data gaps on the Athabasca River are so formidable that "the cumulative effects of water withdrawals on fish and fish habitat in the lower

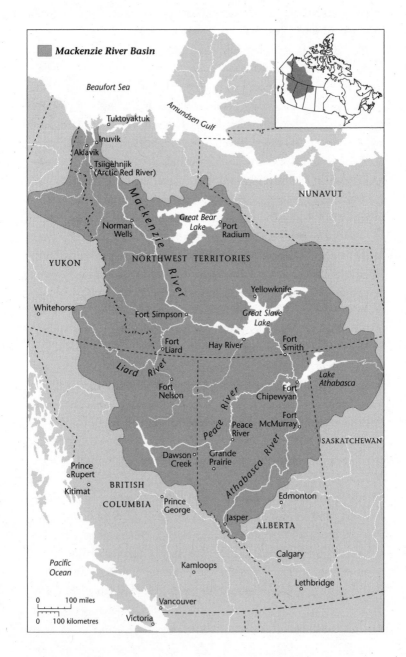

Mackenzie River Basin

Beaufort Sea

Amundsen Gulf

Tuktoyaktuk

Inuvik

Aklavik

Tsiigèhnjik
(Arctic Red River)

NUNAVUT

Norman
Wells

Great Bear
Lake

Port
Radium

YUKON

NORTHWEST TERRITORIES

Mackenzie River

Whitehorse

Yellowknife

Fort Simpson

Great Slave
Lake

Fort
Liard

Hay River

Fort
Smith

Liard River

Lake
Athabasca

Fort
Nelson

Peace River

Fort
Chipewyan

Peace
River

Fort
McMurray

Prince
Rupert

Dawson
Creek

Grande
Prairie

Athabasca River

SASKATCHEWAN

Kitimat

BRITISH

COLUMBIA

Prince
George

Edmonton

Jasper

ALBERTA

Pacific
Ocean

Calgary

Kamloops

Lethbridge

0   100 miles

0   100 kilometres

Vancouver

Victoria

Athabasca River watershed [can't] be predicted with confidence." The World Wildlife Fund warns that warming temperatures "will significantly reduce both water quality and water quantity in the region." Downstream users are already sounding alarm bells about water quality. "Everybody is convinced that the oil sands is having an impact on the basin," says Michael Miltenberger, minister of environment and natural resources for the government of the Northwest Territories. "We have tremendous concerns in terms of the pace of development and contamination issues. What happens on the Athabasca affects people as far away as Inuvik."

The open-pit mines that scar the banks of the Athabasca River north of Fort McMurray are water consumers as formidable as California irrigation projects. Shell's Albian Sands project will not only destroy 31,000 acres of water-conserving forest and wetlands but drink nearly 1.9 billion cubic feet of water a year from the Athabasca River. In addition to trashing 320 acres of fish habitat along the Muskeg and Firebag rivers, Imperial's Kearl project will suck up another 3.7 billion cubic feet from the Athabasca River (2.3 per cent of the river's flow) as well as 317 million cubic feet of groundwater. Kearl will also destroy enough forest, fens, bogs, and wetlands to cover twenty thousand football fields. As Liz Moore's Web site notes, Syncrude takes enough water from the Athabasca River (2.5 trillion gallons) to annually fill the glasses and bathtubs of a third of Denver's residents. In addition to destroying much of the Tar River and its tributaries, the Calumet River and its tributaries, a tributary to the Pierre River, an unnamed tributary to the Athabasca, and an unnamed lake, CNRL's Horizon mine will suck up 3.2 billion cubic feet of water from the Athabasca River. The mine will also reduce groundwater flow into the river by a million cubic feet a day.

Bitumen is one of the most water-intensive hydrocarbons on the planet. (Shale oil is a close second. Colorado shale-oil developments proposed by Shell and Exxon, for example, will use as much water as the tar sands and already threaten what's left of the depleted Colorado River.) On average, the open-pit mines require twelve barrels of water to make one barrel of molasseslike bitumen. Most of the water is needed for a hot-water process (similar to that of a giant washing machine) that separates the hydrocarbons from sand and clay. Although companies such as Syncrude recycle their water as many as eighteen times, every barrel of bitumen consumes a net average of three barrels of potable water. Given that the industry produces one million barrels of bitumen a day, the tar sands industry virtually exports three million barrels of water from the Athabasca River daily.

Industry's water consumption is directly related to the quality of ores recovered from the open-pit mines. According to Bruce Peachey, president of Edmonton-based New Paradigm Engineering and one of Alberta's most clear-headed analysts, industry will actually need more water as it processes dirtier and dirtier bitumen deposits: "We are presently mining the best ores. But as clay content increases, the volume of water needed in production will increase. So this is the good time for water."

The tar sands industry now accounts for more than 76 per cent of the water allocations on the Athabasca River, or 8 per cent of all the water licensed in the province. Current permits allow industry to suck out 2.3 billion barrels of fresh water a year, enough to supply two cities the size of Calgary. (Natural Resources Canada researcher Randy Mikula calculates that's the same amount of water going over Niagara Falls during an eight-hour period.) Planned expansions could bring the total to 3.3 billion barrels per year, a volume that the Natural Resources Canada website admits "would not be sustainable because the Athabasca River does not have sufficient flows."

For nearly a decade, scientists, as well as environmental and Aboriginal groups, have asked the government to study how these city-scale withdrawals are impacting the river's health and instream flows. To date, nobody can say with any certainty whether the province's promiscuous permission-granting has left enough water in the Athabasca for the fish. In the wintertime, water levels drop so low that by 2015 industry will be withdrawing more than 12 per cent of the river's flow.

The job of doing a river health study fell to a dysfunctional multi-stakeholder group called the Cumulative Environmental Management

Association (CEMA) in 2003. CEMA, which bases all its decisions on consensus, honoured neither its 2004 nor its 2005 deadlines for an instream flow report, which didn't surprise anybody. (CEMA also failed to produce a watershed management plan for the Muskeg River before rapid tar sands developments had wiped out half the basin.) One ecologist recently described the group's work with open contempt: "A lot of their studies are absolute shit. Some read like the oil sands is nirvana and everything is a win-win. The fundamental issues have been ignored."

Investing in Our Future, a scathing report on rapid oil sands development by Doug Radke, a former deputy minister of the environment, confirmed the depth of fundamental government neglect in February 2007. The candid document acknowledged that the Alberta government didn't have the capacity to properly review environmental impact assessments on tar sands projects and defined the province's ability to enforce environmental regulations as "inadequate." The study also characterized cumulative-effects planning in the tar sands as "unclear, outdated and incomplete." It concluded that the Athabasca River "may not have sufficient flows to meet the needs of all the planned mining operations and maintain adequate instream flows." Downloading critical government responsibilities such as river health studies to groups such as CEMA, the report warned, could "result in decisions being watered down to the extent that they do not meet the best interests of any party or the environment."

Three months later, Alberta Environment and the federal Department of Fisheries and Oceans finally produced an interim plan, or framework, for the Athabasca River. The plan gives the province until 2010 to figure out if the industry permits it's issued have left enough water in the river to sustain fish. In the meantime, the provincial plan will work like a primitive stoplight. Green-light conditions allow industry to withdraw up to 15 per cent of the Athabasca's flow; a yellow light encourages industry to proceed with caution by reducing water consumption to 10 per cent of flow; and a red light, or fish-killing zone, restricts allocations further. But even during a drought, industry will get enough water to fill fifty bathtubs per second. Bitumen, in other words, comes before fish.

Preston McEachern of Alberta Environment calls the plan conservative and precautionary: "There is a lot of water in that river...and our numbers are low compared to industrialized rivers in Europe or the United States." He says the red-light system will challenge new companies "to plan their projects with exceptional environmental controls." Industry maintains that the river "is not short water." However, Amy Mannix, a Ph.D. student at the University of Alberta, recently calculated that average water levels fell so low between 1999 and 2003 that Alberta's water management framework would have triggered more red and yellow conditions on the beleaguered river every year than green ones. In 2001, for example, low flows would have mandated water restrictions for forty-three weeks. Mannix predicted that the risk of a "water crisis" during a dry year might well drive industry "to put significant pressure on government to ease or remove the restrictions of the framework."

David Schindler, one of the world's foremost water ecologists, describes the province's belated framework as "inadequate." In 2007, he and two other researchers, William Donahue and John P. Thompson, published a reality check on industry's water addiction for the Program on Water Issues at the University of Toronto's Munk Centre and the University of Alberta's Environmental Research and Study Centre.

Their findings were shocking. Net water runoff in the summer and winter months, when the Athabasca River is most vulnerable, had declined by 30 per cent since 1971 due to climate change and could drop to as low as 50 per cent by 2050. "The declines in winter flows are as if someone had added an oil sands plant to the river every two years," the report said.

Schindler, who supports a moratorium on future projects, believes that climate change and industry's escalating demand for water are on a hellish collision course. He reckons that both the Alberta government and Canada's federal government have failed to collect the necessary data for calculating how much water is required to feed the marshes and wetlands of the Athabasca Delta, let alone to keep fish alive. When Schindler delivered these startling findings to a select audience at the University of Alberta in Edmonton in May 2007, neither industry leaders nor the ERCB attended his talk. "It was telling," says Schindler. "Industry is so smug that they can do what they want that they don't want to know what they are doing to the environment."

But not all industry players have been so oblivious. Suncor, the first enterprise in the sands, has reduced its water consumption by 30 per cent in the last two years. One of its facilities uses no fresh water at all. "We are flipping the paradigm from the myth of water abundance to the reality of water scarcity," says Gord Lambert, Suncor's vice-president of sustainable development. "The status quo is not acceptable from an economic and environmental point of view." John Robertson, a senior manager with global engineering giant CH2M Hill, adds, "With pressure on water in northern Alberta, one thing is for sure: the industry will have to spend hundreds of millions in the next few years to treat and reuse water."

In fact, except for members of the Alberta government and one industry group that claims "even with ambitious growth" the Athabasca will remain underutilized, most observers now recognize that current water usage on the Athabasca River is recklessly unsustainable. An October 2007 report on the tar sands by Scotia Capital, aptly titled *Prepare for Glory,* concluded that planned projects would soon consume nearly 9 per cent of the river's winter flow. The report warned that industry probably has "another one to two years before this issue comes to the forefront at which point approvals will become more difficult to obtain, adding a premium to those companies whose projects are pre-approved, or projects that use no water."

In its 2007 review of the oil sands, the Canadian Parliament conservatively called water use by the industry "an enormous challenge... In view of the pace of development being considered, the Athabasca basin could encounter serious problems unless there is a radical change in technology in terms of water use." In 2006, the National Energy Board had come to a similar conclusion: "The limited available supply from the Athabasca River could be a constraint on future expansion plans." A water supply security analysis by Golder Associates, a global engineering firm, concluded that winter levels on the river have hit "extreme low flows" and "new and existing water licenses are at risk."

If water shortages were to occur, both industry and government have limited courses of action. According to Bruce Peachey, companies can either reduce water consumption or build upstream, off-site storage for water taken from the Athasbasca during high spring flows. (After consultation with David Schindler, at least one CEO ordered his company to build an extra thirty-day storage facility.) The government could also limit or shut down bitumen production altogether. Although industry and government have set goals of three million barrels a day by 2015, Peachey thinks water availability could well constrain such exuberance: "The growing public awareness of the need to protect environmental resources and the concurrent needs to protect local communities from the sudden loss of a major employer will create considerable social conflicts over any solution proposed."

Last but not least, lawsuits might start flying. A 2007 article in the *University of Toronto Faculty of Law Review* concluded that if a prolonged drought were to reduce water allocations on the Athabasca River and seriously impact oil production, a U.S. company could construe the cancellation of a water licence as an expropriation under the North American Free Trade Agreement and sue the Alberta government for compensation. "Considering the magnitude of the investment which exists in Alberta's oil sands, as well as the industry's extreme reliance on water for resource recovery, such an example is within the realm of possibility," the article stated. Preston McEachern of Alberta Environment agrees: "Those type of scenarios could be played out in the future and could become a real test of political will."

City-sized open-pit mines will soon be eclipsed by another water hog in the tar sands: in situ production. About 80 per cent of all bitumen deposits lie so deep under the forest that industry must melt them into black syrup with technologies such as steam-assisted gravity drainage (SAGD). Twenty-five SAGD projects worth nearly $80 billion could produce four million barrels of bitumen a day by 2020 and easily surpass mine production. But as Robert Watson, president of Giant Grosmont Petroleum Ltd., warned in 2003 at a regulatory hearing: "David Suzuki is going to have problems with SAGD. Alberta natural gas consumers are going to have problems with SAGD... SAGD is not sustainable."

Watson wasn't kidding. Land leased for SAGD production now covers an area the size of Vancouver Island, which means in situ drilling will threaten water resources over an area fifty times greater than that affected by the mines. SAGD is not benign: it generally industrializes the land and its hydrology with a massive network of well pads, pipelines, seismic lines, and thousands of miles of roads.

Although industry spin doctors calculate that it takes about one barrel of raw water (most from deep salty aquifers) to produce four barrels of bitumen, most SAGD engineers admit to much higher water-to-bitumen ratios. Actually, SAGD could be removing as much water from underground aquifers as the mines are withdrawing from the Athabasca River within a decade. "SAGD is just as big a problem as the mines, and it's not going away. We don't have a plan or strategy for it other than reducing water usage as fast as possible," says Peachey.

Moreover, SAGD's water thirst appears to be expanding. Industry used to think that it only needed two barrels' worth of steam to melt one barrel of bitumen out of deep formations, but the reservoirs have proved uncooperative. Opti-Nexen's multibillion-dollar Long Lake Project south of Fort McMurray, for example, originally predicted an average steam-oil ratio of 2.4. But Nexen now forecasts a 35 per cent increase in steam (a 3.3 ratio). Most SAGD projects have increased their steam ratios to greater than three barrels, with a few projects already as high as seven or eight. "A lot of projects may prove uneconomic in their second or third phases because it takes too much steam to recover the oil," explains one Calgary-based SAGD developer.

High-pressure steam injection into bitumen formations can cause micro earthquakes and heave the surface of land by up to eight inches. Steam stress can also fracture overlying rock, allowing steam to escape into groundwater or the empty chambers of old SAGD operations. (The steam stress problem is so dramatic, says one engineer, that all forecasts of SAGD potential production are probably grossly exaggerated.) Both Imperial Oil and Total have experienced spectacular SAGD failures that left millions of dollars of equipment soaking in mud bogs.

The dramatic loss in steam efficiency for deep bitumen deposits means companies have to drain more aquifers to boil more water. To boil more water, the companies have to use more natural gas (the industry currently burns enough gas every day to keep the population of Colorado warm), which in turn means more greenhouse gas emissions. By some estimates, SAGD could consume the entire gas supply of Western Canada by 2025.

SAGD's frightful natural gas addiction is now driving shallow drilling as well as coal-bed methane developments on prime agricultural land throughout central Alberta. (Coal-bed methane is the tar sands of natural gas: it requires more wells and more land disturbance than conventional gas and poses a huge threat to groundwater, which often moves along coal seams.) The quick removal of natural gas from underground pools and coal deposits creates a void that could, over time, fill up with either water or migrating gas. Nobody really knows at the moment how many old gas pools connect with water aquifers or how many are filling up with water. Bruce Peachey estimates that natural gas drilling could result in the eventual disappearance of 350 to 530 billion cubic feet of water in arid central Alberta. That's enough water to sustain the city of Beijing with its population of fifteen million people for three years.

Due to spectacular growth in SAGD (nearly $4 billion worth of construction a year until 2015), Alberta Environment can no longer accurately predict industry's water needs. The Pembina Institute, a Calgary-based energy watchdog, reported that the use of fresh water for SAGD in 2004 increased three times faster than the government forecast of 110 million cubic feet a year. Government has made a conscious effort to get SAGD operations to switch to using salty groundwater. However, since it costs more to desalinate the water and creates a salt disposal problem, SAGD could be still be drawing more than 50 per cent of its volume from freshwater sources by 2015. (SAGD now accounts for 7 per cent of the water withdrawals from the Athabasca River.) And even at a 50 per cent use of groundwater, SAGD generates formidable piles of toxic waste. Companies can't make steam without first taking the salt and minerals out of brackish water. As a consequence, an average SAGD producer can generate thirty-three million pounds of salts and water-solvent carcinogens a year, which simply gets trucked to landfills. Because the waste could contaminate potable groundwater, industry calls its salt disposal problem "a perpetual care issue." Even the U.S. Congressional Research Service calls SAGD waste "a serious disposal problem."

To reduce its annual $26-million waste liabilities, Opti-Nexen recently purchased a 190,000-pound water treatment system in the United States. The company hired a 230-foot truck trailer, one of the longest in North America, to import the equipment. Yet insiders remain alarmed by industry's rising salt budget. "There is no regulatory oversight of these landfills, and these problems will be enormously difficult to fix," says one SAGD developer.

Arsenic, a potent cancer-maker, poses another challenge. Industry acknowledges that in situ production (the terrestrial equivalent of heating up the ocean) can warm groundwater and thereby liberate arsenic and other heavy metals from deep sediments. Canadian Natural Resources recently reported that one arsenic plume moved nearly twelve hundred feet over a fifteen-year period but estimated "it would take centuries, if ever," for the arsenic to affect drinking water. No one, however, knows how much arsenic seventy-eight approved SAGD projects will eventually mobilize into Alberta's groundwater and from there into the Athabasca River.

The biggest issue for SAGD production may be changes in the water table over time. "If you take out a barrel of oil from underground, it will be replaced with a barrel of water from somewhere," explains Bruce Peachey. The same rule applies to natural gas. Peachey figures that if all the depleted gas pools near the tar sands were to refill with water, the water debt could amount to half the Athabasca River's annual flow. This vacuum effect may also explain why the most heavily drilled energy states in the United States are experiencing the most critical water shortages.

The paucity of groundwater data in the Athabasca region has been an outstanding concern for years. In 2007, the Rosenberg International Forum on Water Policy, a prestigious water group at the University of Calgary, warned the Alberta government that water was "every bit as important as oil" and said that the existing network of groundwater monitoring was "insufficient to provide reliable information on water quality and water levels." The group also revealed that resource development had outstripped the pace of scientific research.

Bitumen production has sparked explosive confrontations on this very issue with cottagers in Alberta's northern lake country. In 2007, Don Savard, the retired senior vice-president of Enbridge Pipelines, learned that a Calgary firm wanted to mine billions of barrels of bitumen under his cabin on Marie Lake, 112 miles north of Edmonton. The lake, a jewel of clarity and quality in a province with fewer than eight hundred lakes, had successfully withstood increasing levels of SAGD tar sands development for twenty years. But the Calgary firm wanted to use unproven tunnel technology and to map the deposit with thousands of sonic blasts from floating air cannons. Savard didn't think the lake's fish or ecosystem could survive the assault, so he picked up his phone. Hundreds of cottagers and local landowners mounted a battle for Marie Lake that became a cause célèbre in water-short Alberta. In June 2007, Premier Ed Stelmach abruptly cancelled the company's development lease under the lake. Savard still doesn't know how or why his group won. But for the oil-patch veteran, the controversy personified the conflicted state of water management in a petrostate. "We are producing oil for somebody else and overburdening the economy and affecting our quality of life...Somewhere along the line, and we're not doing a very good job of it, we need to say no. We need to protect our lakes and rivers."

ALBERTA ENVIRONMENT PROUDLY claims on its Web site that it does not proceed with development "at the expense of the environment." To make its case, the department points to annual reports by the Regional

Aquatics Monitoring Program (RAMP), a quasi-private stakeholders group set up by industry and government in 1997. Indefatigable RAMP has consistently reported the kind of fairy tale that industry and government spin doctors appreciate: rapid tar sands development has never dirtied the Athabasca River. The government's repeated assertions of "no significant impacts" show chutzpah. Albertans are expected to believe that the world's largest energy project can displace more than a million tons of boreal forest a day, industrialize a landscape mostly covered by wetlands, create twenty-three square miles of toxic-waste ponds, spew tons of acidic emissions, and drain as much water from the Athabasca River as that annually used by Toronto, all with no measurable impact on water quality or fish.

This surreal claim owes much to the work of RAMP. Although the Alberta government calls RAMP a "community group," it receives its funding from a select community: Syncrude Canada Ltd., Suncor Energy Inc., Albian Sands Energy Inc., Shell Canada Ltd., Canadian Natural Resources Ltd., Imperial Oil Resources, Petro-Canada Oil and Gas, OPTI Canada Inc./Nexen Inc., Husky Energy, and Total E&P Canada Ltd. Since its inception, RAMP has collected and reported data on water quality, climate, hydrology, and fish and invertebrate (snail and clam) health for a 26,0000-square-mile region. In 1999, RAMP added fifty lakes to its monitoring program, ostensibly to keep track of great clouds of acidic emissions from the upgraders. Every RAMP report, though few Albertans or U.S. gasoline buyers have seen one, makes for a pleasant reading experience. Bitumen production and its sulfurous emissions apparently do not dirty water in Alberta. Or harm fish. Or acidify lakes.

RAMP's first report, in 1997, found "no increases in concentrations of parameters" or pollution from tar sands activity. Nor did the study's sampling find "consistent evidence of an influence on oil sands operations on wetlands or associated plant communities." After thirty years of open-pit mining, leaking tailings ponds, and numerous oil spills from two of the world's largest open-pit mines, the Athabasca River supposedly remained in great shape.

The following year, RAMP reported distressingly low water levels but overall declared the Athabasca River to be "non-toxic." In 1999, water quality remained "within historic trends." A year later, despite "a large increase in oil sands development," the good news continued: "Water and sediment quality in the Athabasca River and tributaries to the Athabasca River was generally consistent with historical data." The 2001 report brought more positive salutations: "no significant trends in water quality over time" had occurred, even though ten companies now operated or were building seventeen projects along the river. A RAMP Five-Year Report Summary (1997–2001) presented more auspicious data: "the increase in development downstream of Fort McMurray has not degraded the water quality." In 2003, water quality "was generally consistent with previous years," and the fifty lakes monitored for evidence of acid sensitivity reportedly all fell within the normal range. By 2004, ten multinationals were building twenty-one tar sands projects, but "any influences on the Athabasca River" appeared to be minor. And on it went.

Three prominent federal civil servants including G. Burton Ayles, then director general, Central and Arctic Region, Department of Fisheries and Oceans in Winnipeg, finally explained in 2004 why so much industrial activity in a boreal watershed amounted to "no significant effects." With the help of fifteen expert researchers, the three senior scientists highlighted in their detailed peer review a raft of "serious concerns" with the quality of RAMP's monitoring, due to a chronic lack of scientific leadership, effective design, and consistent monitoring.

The experts who reviewed RAMP's work on climate and hydrology, for example, found no design and no plan. They expressed disbelief at "the modest attention given the Athabasca River system...particularly in light of the lack of long-term data sets for the reaches downstream of oil sand developments." After concluding that water sampling "may be inadequate for detecting effects, especially cumulative effects," they strongly recommended "a monitoring design and analytical plan." They also demanded that RAMP immediately establish regional groundwater monitoring—something that did not begin until 2007.

The scientists who reviewed the water quality data were equally critical. They concluded that the number of monitoring sites was inadequate (there was just one station downstream from the tar sands) and that the monitoring program had sampled so many different things from so many different places over time that the data couldn't measure, let alone detect, any kind of impact. They rated RAMP's wishful conclusions of no change in water quality as "not warranted" and expressed alarm "that the main monitoring program for the area significantly lacks strategic direction and scientific process." Furthermore, the scientists explained, RAMP's water quality program was not designed to register "development-related change locally or in a cumulative way."

Snail and clam experts concluded the present sampling protocols were almost useless: "It is clear that a suitable, overall effects-based monitoring design must be adopted, or development related change will not be assessed." They called for the Athabasca River to be included "in any monitoring program for oil sands development." The fish experts in the peer review said that testing for different chemicals in different species from different sites in different years added up to a program that "lacks a clear focus." The vegetation experts found RAMP sampling to be haphazard at best and recommended "less representation by industry and more representation by nonpartisan groups...the make-up of the RAMP committees is too heavily weighted towards industry." As for the acid-sensitive lakes program, experts found it had been designed so poorly and with so little baseline and historical data that it was "unlikely to achieve its stated objectives."

In a 2008 e-mail to David Schindler, Burton Ayles, one of the review's three authors, said the critiques had changed nothing at RAMP: "I am pretty sure there was no systematic assessment of the report, i.e., they never said 'This looks good, let's set up a subgroup to make sure these recommendations are carried out.' We, of course, were hoping to be able to say, 'Look how we turned things around with a complex environmental program.' No such luck I am afraid. They did change contractors but not much else seemed to happen." Schindler adds that if a group of peer reviewers had made the same comments on a federal monitoring program, "it would have been canned. It's that simple."

David Rosenberg, a private consultant, former environmental scientist with the Department of Fisheries and Oceans, and one of the peer reviewers, agrees wholeheartedly with Schindler. "We gave them an analysis they didn't want, and they never came back to us."

The findings regarding RAMP's fraudulent design did not surprise Rosenberg. Industrial groups such as RAMP repeatedly design their studies to "find the project acceptable" regardless of the scientific evidence, says Rosenberg. "The fact that they went on blithely and didn't use the powerful biomonitoring technologies we recommended in our report is really stunning." Rosenberg believes that "fresh" and "water" combined are dirty words in Ottawa, and he notes that while the United States knows the amount of water running in every one of its streams, "up here we are flying blind." He sees "no political will" to fix the pattern of national water abuse.

Only a nation without a water policy could allow such rapid development of the tar sands in the world's third-largest water basin. (Only the Amazon and the Mississippi beat out the Mackenzie watershed in size.) Canada is not only without a water strategy; it collects little groundwater data and has one of the industrial world's poorest freshwater monitoring programs. (Northern Canada has only thirty-six monitoring sites.) It's not as though Ottawa doesn't understand the scale of its neglect. Classified government documents obtained by Ottawa public-interest researcher Ken Rubin confirm "that the absence of a common strategic federal vision for freshwater" is a "limiting factor for ensuring the long-term sustainable development of the resource." Environment Canada systematically fails to uphold federal laws against pulp-mill polluters in Atlantic Canada and the hog industry in Manitoba and Quebec. As a consequence, both Lake Winnipeg and the St. Lawrence River are on the verge of ecological collapse. Statistics Canada reported in 2007 on the nation's Third World water state: 23 per cent of Canada's waterways can no longer support aquatic life, due to phosphorous, nitrogen, and ammonia pollution.

Alberta's water record is just as dismal. After overallocating resources in every major watershed in drought-prone southern Alberta, the province reluctantly closed the South Saskatchewan River basin three years ago, due to fish kills and declining river health. It seems poised to repeat the same mistake on the Athabasca.

Persistent declines in water quantity and quality are compounded by a shortage of basic information on surface water and groundwater. A 2008 report by the Alberta Water Council, a nonprofit watchdog set up by the province, described "the availability, quality and accessibility of data" as a real problem. The report found that 245 agencies and citizens responding to a survey felt the province's aquatic environment was being "slowly and steadily compromised." Brad Stelfox, a prominent land-use ecologist who works for both industry and government, notes that a century ago all water in Alberta was drinkable. "Three generations later all water is non-potable and must be chemically treated," he points out. "Is that sustainable?"

The water-data gap has also raised alarm bells at the Petroleum Technology Alliance of Canada. The alliance regards water quantity and quality as "the environmental issue of the century." In a recent paper, it concluded that "rapidly growing demands for water, where data is limited due to reduced government-supported data gathering in the last 20 to 40 years, will drive and limit development."

A year after the disastrous peer review, RAMP commissioned an investigation of its flawed program design. Examiners reviewed the predicted environmental impacts for seventeen mines and SAGD projects, then checked to see if the RAMP process had registered any actual changes. Altogether, the projects studied affected 222 overlapping watercourses and water bodies, or an average of 13 creeks, streams, and rivers per project.

The commissioned report confirmed the conclusions of the 2004 peer review by explaining in greater detail why RAMP never found any bad news. The reason was simple: RAMP's current design made it "difficult" to measure damage at all, and it wasn't much good beyond registering the impact of a single project at a single site. The program's ability to detect regional effects was "uncertain"; the report even suggested that "current RAMP design provides data that are in fact not suitable for monitoring regional trends." All in all, the report concluded, "the recorded data does not have sufficient accuracy in many cases to distinguish between negligible and low impacts or even low and moderate impacts."

Nor could RAMP's program detect impacts on climate, hydrology, or fish populations, because "generally accepted and approved effects criteria do not exist." The report also noted that RAMP's failure to collect reliable baseline data for a period of at least two decades prior to rapid development undermined the integrity and reliability of the monitoring. Furthermore, "the number of monitored watersheds that are unlikely to be affected by future development is very small and is decreasing as new development plans are announced."

Undaunted by these failing grades, RAMP continues to pump out reports on water quality that largely reinforce the program's myth-making function. In its eleventh "community report" in 2007, RAMP predictably declared to the eighty thousand residents of Wood Buffalo, the hundred thousand northern Canadians who live downstream from the tar sands, and millions of tar sands investors that "there were no detectable regional changes in aquatic resources related to oil sands development." Politicians cite RAMP's Disney-like declarations with abandon. In June 2008, Member of Parliament Brian Jean (Fort McMurray–Athabasca) offered RAMP's misleading conclusions to a somewhat skeptical Standing Committee on Environment and Sustainable Development with the enthusiasm of a Boy Scout.

"So there's been no change," he told the committee. ". . . In fact they also found in this report that 'no effects of local human actitivities were apparent on the quality in the Athabasca River in 2007.' Were you aware of that?"

SIX

THE PONDS

....

"Syncrude offers a heartfelt and sincere apology for the incident
on April 28th that caused hundreds of migratory birds to die after
they landed on a tailings pond at our oil sands operation."

TOM KATINAS, SYNCRUDE PRESIDENT AND CEO, ADVERTISEMENT
IN THE *NATIONAL POST*, 2008

IN THE SPRING of 2008, five hundred ducks made international news by
landing on Syncrude's Aurora North Settling Basin, which locals now
call Dead Duck Lake. It's a large body of toxic waste covered with bitumen as sticky as Krazy Glue. Many of the migrating visitors were
buffleheads, keen divers that slipped under the water and never resurfaced. When Syncrude managers failed to report the incident, a company
whistle-blower alerted authorities and Greenpeace. Before anyone
could say quack, Alberta Premier Ed Stelmach and Canadian Prime
Minister Stephen Harper had apologized for the tragedy. Stelmach
called the deaths an opportunity to prove the province's profound commitment to environmental justice. Alberta's environment minister, Rob
Renner, a florist by trade, said the news saddened him. Syncrude's CEO
Tom Katinas honoured the dead ducks in full-page newspaper ads,
promising that "a sad event like this" would never happen again. In an

operatic attempt to salvage Alberta's dirty oil reputation, the provincial government airlifted some well-tarred ducks to Edmonton for special cleaning. Most died. No one apologized for the tar sands'—and Canada's—greatest, most cancerous liability: the rapid growth of twenty-three square miles of leaking toxic ponds along the Athabasca River.

Astronauts can see the ponds from space, and politicians typically confuse them with lakes. Miners call the watery mess "tailings." Industry prefers the term "oil sands process materials" (OSPM). Call them what you like, there is no denying that the world's biggest energy project has spawned one of the world's most fantastic concentrations of toxic waste, producing enough sludge every day (400 million gallons) to fill 720 Olympic pools.

The sheer scale of these ponds invites disbelief. Engineers proudly describe the man-made dams as "some of the biggest structures in the world." The Alberta Chamber of Resources, an industry cheerleader, calls the primitive storage system "a risk to the oil sands industry," while the National Energy Board, a federal promoter of rapid energy development, describes the ponds merely as "daunting." The Canadian Parliament, a polite entity, refers to the impoundments of waste as "enormous challenges to the industry and researchers" alike. The Alberta Energy Research Institute, an agency not known for its environmental radicalism, calls the exponential growth of the ponds "unsustainable."

The ponds are truly a wonder of geotechnical engineering. Made from earth stripped off the top of open-pit mines, they rise an average of 270 feet above the forest floor like strange flat-topped pyramids. By now, the ponds hold more than four decades' worth of contaminated water, sand, and bitumen. Amazingly, regulators have allowed industry to build nearly a dozen of them on either side of the Athabasca River. The river, as noted, feeds the Mackenzie River Basin, which carries a fifth of Canada's fresh water to the Arctic Ocean. The basin ferries wastes from the tar sands to the Arctic too.

The ponds are a byproduct of bad design and industry's profligate water abuse. Of the twelve barrels of water needed to make one barrel of bitumen, approximately three barrels become mudlike tailings. All in all, approximately 90 per cent of the fresh water withdrawn from the Athabasca River ends up in settling ponds engineered by firms such as Klohn Crippen Berger and owned by the likes of Syncrude, Imperial, Shell, or CNRL. After separating bitumen from sand with hot water and caustic soda, industry pumps the leftover ketchuplike mess into the ponds.

Engineers originally thought that the clay and solids would quickly settle out from the water in the ponds. But bitumen's clay chemistry confounded their expectations, and the ponds have been stubbornly growing ever since. They now cover twenty-three square miles of forest and muskeg. That's equivalent to nearly 120 Moraine Lakes, the pretty body of water that appeared on Canada's old twenty-dollar bill, or more than 75 Lake Louises without the Rocky Mountain scenery. Within a decade, the ponds will cover an area of eighty-five square miles. Experts now say that it might take a thousand years for the clay in the dirty water to settle out.

Alberta Environment, a reliable bearer of good news, says that the ponds contain only "stable dispersions of bitumen, clay and water," and so are safe. Natural Resources Canada offers a more complete list of contaminants, stating that salt, phenols, benzene, cyanide, heavy metals (such as arsenic), and dozens of other cancer makers can be found in the ponds. The impoundments smell like a leaky gas station due to all these volatile compounds and freeze only in the coldest weather. Minnows dropped in their waters die within ninety-six hours. As Syncrude's disaster demonstrated, unwary ducks get coated in bitumen and sink to the bottom.

Stephen Borsy, a Calgary-based tar sands worker, described the ponds as an environmental "nightmare" in a vivid 2006 Internet posting. "How these companies can say that [the ponds] are environmentally friendly I'll never understand. Just because they reclaim some of the land after strip-mining, then plant some grass and trees, and put in a few wood-bison to roam there, does not mean that they are doing anything that is environmentally friendly. The tailings are a soup of wastewater crap that literally sits in a pool called a tailings pond. Even at −30°C, this stuff doesn't freeze. It just sits there, steaming. The stench from these ponds is indescribable. I've been there. I've seen it. I've smelled it."

Scientists have pegged acutely toxic tailings waste as an issue from day one because companies could not legally dump the stuff into the river. In 1977, a Syncrude researcher described the "sludge problem" as the "single most serious problem to be faced in an oil sands mining operation." Nearly a decade later, Environment Canada admitted that "the accumulation of tailings pond sludge and toxic pond water may pose an environmental problem for many years to come."

Randy Mikula, team leader for Natural Resources Canada's CANMET Energy Technology Centre in Devon, Alberta, has been studying the ponds for twenty-two years. He calls their continued expansion both "frightening and vexatious." Mikula calculates that industry to date has created seven billion cubic yards of sandy gunk, of which one billion cubic yards are fine tailings, enough toxic waste to fill a canal thirty-two feet deep and sixty-five feet wide from Fort McMurray to Edmonton and on to Ottawa. "It's not just an Alberta problem."

Jim Byrne, a water expert at the University of Lethbridge, recently offered a similarly startling equation: if Alberta drained its tar sands waste into Lake Erie, it would fill the basin to a depth of eight inches today. By 2030, this toxic soup would be nearly seven feet deep. Not many nations other than China can claim such records of waste production.

Like most scientists working in the tar sands, Randy Mikula regards the scale of the ponds as "mind boggling." CNRL's Sand Starter Dyke for the Horizon Oil Sands mine will cover 8,850 acres with a top height of 210 feet and will eventually house 35 billion cubic feet of gunk. Other ponds range in size from 370 acres to 7,500 acres. Suncor's new South Tailings Pond encompasses six square miles of forest and will hold 12 billion cubic feet of waste. And so on.

Syncrude, the largest oil producer in the tar sands, owns the biggest pond of all. Every day, the company dumps 500,000 tons of tailings into the Syncrude Tailings Dam, which the U.S. Department of the Interior officially rates as the world's largest by volume of construction material. The pond, completed in 1973, stretches for fourteen miles and holds 19 billion cubic feet of water, crud, and sand. When China completes the Three Gorges Dam in 2009 (it's already submerged 13 major cities, 140 towns, and 326 peasant villages) Syncrude's dam will become, as Mikula notes, "just second best."

Engineers such as Dr. Gord McKenna at Syncrude admit that they built these ponds with "challenging materials" such as sand and clay on "generally some of the weakest foundation conditions found anywhere," including muskeg. In fact, Syncrude's Geotechnical Review Board has tackled about every mess possible during the construction of forty-six square miles of earth dumps and tailings ponds north of Fort McMurray. In a 1998 issue of the *Geotechnical News*, the board offered a list of mistakes or "challenges" that included the depressurization of aquifers; the failure of early dykes; constant leaking; the movement of dyke foundations; the collapse of walls; frost effects on foundations; a progressive failure in sands and clays; and the instability of reclaimed landscapes, to name a few.

Every year the ponds quietly swallow thousands of ducks, geese, and shorebirds as well as moose, deer, and beaver. Although Canada's Migratory Birds Convention Act says it's against the law to kill birds by sliming them with bitumen or other toxic waste, shit happens in the tar sands. (In response to the recent five-hundred-duck fiasco, Environment Canada is considering issuing permits to tar sands companies to make the drownings legal.) To a bird's eye, the toxic ponds look like a nice bit of ice-free real estate in the spring and fall. The ponds also lie under a major migratory flyway for birds travelling to the Peace-Athabasca Delta, which Environment Canada calls "one of the most important waterfowl nesting and staging areas in North America." A trumpeter swan or snow goose coated in bitumen often dies of hypothermia. Dene and Métis hunters have found slimed, near-dead ducks 135 miles north of the ponds in the bush.

Industry has tried to keep bird killing to a minimum by using scarecrows affectionately called Bit-U-Men. Some companies also fire propane cannons to keep ducks, geese, and shorebirds off their deadly waters. The birds, however, are used to the racket. Albian Sands, which just built its own toxic lake, has developed a new "on demand radar based bird deterrent system." Under this novel scheme, flying birds activate a radar system that sets off a combination of deterrents, including scare cannons, robotic falcons powered by solar panels, flashing lights, and recorded attack calls. This multi-alarm system seems to be about five times more effective than scarecrows alone. But the best deterrent is not to make toxic tailings ponds in the first place, says University of Alberta biology researcher Colleen Cassady St. Clair. She notes, however, that technologies to replace the ponds are "likely to be at least 10 years away."

Miners and engineers generally don't canoe on or fish in the ponds because of two really nasty pollutants: polycyclic aromatic hydrocarbons (PAH) and naphthenic acids. Of twenty-five PAHs studied by the U.S. Environmental Protection Agency (and there are hundreds), fourteen are proven human carcinogens. The EPA found that many PAHs produce skin cancers in "practically all animal species tested." Fish exposed to PAHs typically show "fin erosion, liver abnormalities, cataracts, and immune system impairments leading to increased susceptibility to disease." Even the Canadian Association of Petroleum Producers recognizes that a "significant increase in processing of heavy oil and tar sands in Western Canada in recent years has led to the rising concerns on worker exposure to polycyclic aromatic hydrocarbons."

In 2003, the ubiquitous presence of PAHs in the tar ponds prompted entomologist Dr. Jan Ciborowski to make another one of those unbelievable tar sands calculations: he estimated that it would take seven million years for the local midge and black fly populations to metabolize all of the industry's cancer makers.

Naphthenic acids, which by weight compose 2 per cent of bitumen deposits in the Athabasca region, are not much friendlier than PAHs. Industry typically recovers these acids from oil to make wood preservatives or fungicides and flame retardants for textiles. The acids are also one of the key ingredients used in napalm bombs. Naphthenic acids kill fish and most aquatic life. A dosage just ten times greater than what industry expects to leave on a reclaimed tar sands wetland can ruin a rodent's sex life and fry his liver.

The ponds, some locals have discovered, are leaking PAHs and naphthenic acids into the Athabasca River in quantities as great as eighteen gallons a second. John Semple, a Fort McMurray outfitter and longtime resident of the boomtown, loves the river. He is not opposed to the mines, but he doesn't understand why the government has approved one project after another with little regard for people or fresh water. Just outside the Suncor site, where the trees disappear and hydrocarbons fill the air, Semple killed his boat's motor and pointed to an odd-looking cone on the west bank of the river. He explained it had once been an island. Local Cree and Métis called the place Tar Island, because bitumen often oozed down its banks. In the late 1960s, Suncor transformed the island into a tailings pond, the first in the tar sands. To this day, Semple can't figure why Environment Canada allowed the construction of such an impoundment so close to the Athabasca River. "There is stuff coming off there [the dyke] into the river." The ponds leak so routinely, in fact, that they are surrounded by medieval-looking moats equipped with pumping stations to return the seepage to the ponds.

Tar Island Dyke began as an experiment in how to contain bitumen waste. Engineers not only miscalculated how long they would need the pond (their initial projection was a few years) but underestimated the fluidity and instability of the tailings. A dam originally designed to be 40 feet high now towers over the river at 325 feet and stretches for two miles.

Over the years, the dyke has experienced lots of challenging issues, including something called "deformation creep." Dr. Norbert Morgenstern, a highly respected geotechnical engineer and tailings pond expert, explains that creep is movement in a dam's foundations. To stop the creep, Suncor recently installed a small berm at the toe of the dyke and is removing the waste to another pond. According to Morgenstern, the Tar Island Dyke reflects "incremental learning in the industry." The original dyke drained toxic waste directly into the river. Now, as

John Semple suspected, toxins seep into the Athabasca River from the bottom of the dyke at the rate of one million gallons per day. According to a 2007 report prepared for Suncor, that's enough toxic waste to fill about two Olympic-sized pools every day.

In 2001, Morgenstern wrote a powerful paper on tailings ponds in the global mining industry for the European Union. He concluded that many have failed, that their reliability is "among the lowest of earth structures," and that "a well-intentioned corporation employing apparently well qualified consultants is not adequate insurance against serious incidents."

Morgenstern, who sits on Alberta's dam safety committee, didn't think any of his conclusions applied to Alberta's ponds, but others aren't quite so confident. At a 2005 conference on geotechnical engineering for disaster mitigation, two Iranian engineers presented a paper on the Syncrude Tailings Dam and concluded that "the tailings dam and foundation are comparatively in a more critical condition with respect to yield zones, displacements and strains" than expected. Other engineers have expressed worries about the "significant design challenges" posed by building raised toxic lakes on top of large Pleistocene glacial meltwater channel deposits, a common practice in the tar sands. At a 2006 Remediation Technologies symposium, four Canadian engineers revealed that one tar sands pond was leaking naphthenic acids, trace metals, and ammonium into groundwater.

Perhaps the biggest environmental risk is an accidental breach. Earthquakes and extreme weather events can make a rubble of even the best-engineered dykes and could cause the domino-like failure of other nearby ponds. In 2003, the intergovernmental Mackenzie River Basin Board identified the tailings ponds as a singular hazard. The board noted that "an accident related to the failure of one of the oil sands tailings ponds could have a catastrophic impact on the aquatic ecosystem of the Mackenzie River Basin."

Such catastrophes have happened before. In 2000, a tailings pond operated by the Australian-Romanian company Aurul S.A. broke after a heavy rain in Baia Mare, Romania. The pond released enough cyanide-laced water to potentially kill one billion people, and the drinking supplies for twenty-three municipalities were shut down. According to Janos Toth, a Hungarian academic, "The 150 km long toxic tide traveled a distance of 1950 km, flowing through Romania, Hungary, Yugoslavia, Bulgaria, and on into the Black Sea, devastating 1000 km of river ecosystems. According to the damage assessments, 1240 tons of fish were killed."

Engineers and ecologists agree that the tailings ponds pose a substantial risk to Canada's largest river basin. "The longer the tailings sit there, the more likely there will be a major extreme weather event and a big dyke failure," predicts Bruce Peachey of New Paradigm Engineering. "If any of those [tailings ponds] were ever to breach and discharge into the river, the world would forever forget about the *Exxon Valdez*," adds the University of Alberta's David Schindler. (The *Valdez* released about eleven million gallons of crude oil into Prince William Sound, Alaska, in 1989. PAH concentrations alone in the tar ponds represent about three thousand *Valdez*es.)

For now, leaks from the ponds remain a constant challenge. "We know they leak, and we capture these leakages or let some fall into poor quality water formations," says Preston McEachern of Alberta Environment. But when researcher Dr. Kevin Timoney, preparing a health study for the Nunee Health Authority of Fort Chipewyan, asked the government what the actual seepage rate was, McEachern told him that Alberta Environment's procedure was to direct interested parties to contact the companies directly. Timoney considered McEachern's answer essentially "an admission that the government does not know the overall seepage rate."

Most tar sands tailings ponds now seep so badly that they've created toxic wetlands near their bases. Some companies have extended experimental wetlands near these leaks to gather reclamation data. Cattails and hummock grass thrive in the effluent. But a 1999 study in the journal *Ecological Applications* found that indigenous fish were "unable to survive in the wetlands containing oil sands effluent." Tadpoles placed in the tar sands wetlands either expired or grew slowly.

Amphibians, many species of which face worldwide extinction, are sensitive indicators of water quality.

Plant studies have also yielded discouraging results. Tar sands wetlands inhibit the germination of tomatoes, clover, wheat, rye, peas, canary grass, and loblolly pine. Concluded the authors of the 2002 study in *Environmental Pollution*: "The negative effects of the effluents on seed germination may account for the paucity of aquatic species that have invaded the oil sands impacted wetlands."

Birds haven't fared much better. One 2006 University of Saskatchewan thesis looked at how tree swallows coped in a normal wetland versus one made from tar sands tailings waste. The tree swallow, which can nest in a box, makes "an effective indicator species of environmental health." During a bout of harsh weather, mortality rates reached 48 per cent among swallows in a normal marsh; the death rate ranged between 59 per cent and 100 per cent in the tarry reclaimed wetlands. Swallows in the polluted wetlands also showed high levels of thyroid dysfunction, which researchers said could affect their chances of migrating successfully, and the birds seemed to be overwhelmed by blowflies, suffering "parasitic burdens approximately twice those" occurring in natural wetlands. The scientists concluded that nestlings from wetlands made of tailings pond waste "may be less able to withstand additional stressors, which could decrease their chances of survival after fledging."

Seepage is not a short-term problem. When Suncor proposed building its South Tailings Pond to accommodate waste from its Millennium Mine in 2004, its engineers told the ERCB that the 4,700-acre site would contain 12 billion cubic feet of waste, some of which they admitted would seep. Moreover, the engineers said, seepage into groundwater could "change water quality in the lower portion of McLean Creek and could therefore impact the health of aquatic life, terrestrial wildlife and humans." To prevent groundwater contamination, Suncor devised a fancy reverse-pumping system that put contaminants back into its pond. But even after the mine closes, Suncor engineers predicted, the company will have to maintain a "seepage interception system" for sixty years or longer.

Industry has tried all sorts of chemical tricks to get the ponds to settle more quickly and lose their toxicity. They've freeze-dried tailings, added bacteria to the ponds, and even poured in chemical stabilizers such as gypsum. But nothing really works well. Many companies now hope to bury their problem in something called "end-pit lakes." The process involves piping tailings into an old mine site, topping the gunk with millions of barrels of fresh water from the Athabasca River and then waiting for Mother Nature to come up with a solution. Although the region could contain 230 square miles of end-pit lakes by 2040, Alberta's energy regulator admits end-pit lakes are "a complex and as yet unproved concept." Randy Mikula says there is little scientific evidence suggesting the cheap waste-disposal scheme will work.

Fred McDonald, a Métis trapper and storyteller extraordinaire, often questioned the reasoning and science behind the proliferation of toxic ponds and end-pit lakes. Before he died in 2007 of kidney failure, McDonald lived in Fort McKay, an Aboriginal community forty-five miles north of Fort McMurray. The stench of hydrocarbons from the surrounding mines often hangs heavily in the air there, and an ammonia release from a Syncrude facility in 2006 hospitalized more than twenty children.

On a fall day in 2006, McDonald sat in his kitchen, sipping a glass of rat root juice ("It's good for everything," he told me) and breathing through an oxygen tube. The day before, he had spent several hours on a dialysis machine. McDonald's kidneys were failing, but not his mind. He recalled the days when Tar Island was a good place to fish and hunt. "It always had moose on it. We loved that island. We are slowly losing everything."

McDonald was born on the river, and he had trapped, fished, farmed, and worked for the oil companies. He fondly remembered the 1930s and 1940s, when Syrian fur traders exchanged pots and pans for muskrat and beaver furs along the Athabasca River. Families lived off the land then and had feasts of rabbit. They netted jackfish, pickerel, and whitefish all winter long. "Everyone walked or paddled, and the people were healthy," McDonald said. "No one travels that river anymore. There is nothing in that river. It's polluted. Once you could dip your cup and have a nice cold drink from that river, and now you can't."

McDonald said that tar sands pollution is killing berries. The mines are also draining the surrounding muskeg of water: "It's our future source of water, and it's drying." Climate warming has changed the clear blue ice of the Athabasca River in the winter to a dangerous slush. McDonald had recently told his son not to have any more children: "They are going to suffer. They are going to have a tough time to breathe and will have nothing to drink." He dismissed the talk of reclaiming waste ponds and open-pit mines as a white-skinned fairy tale. "There is no way in this world that you can put Mother Earth back like it was."

Because of "the bad behaviour of clays," Randy Mikula suspects that tar sands waste won't settle to solid form for a thousand years, so "something has to be done." Right now the best solution might be a "brute force" centrifugal approach, says Mikula. Waste is spun (much like lettuce in a spinner) in order to create material that is dry and stackable, while recovering water at the same time. Both Syncrude and Suncor have started pilot projects. "We could reduce water usage by a barrel, which means less water withdrawn from the Athabasca River," Mikula says.

The volume of sand and toxic waste produced by the tar sands to date is as great as the agricultural drainage and sewage water the water-short nation of Egypt, with a population of eighty million, reuses every year. By 2015, the tar sands may have created ponds of wastewater three times that size.

THE GROWING WASTE problem is nowhere more evident than downstream in Fort Chipewyan, where the Athabasca and Peace rivers spill into Lake Athabasca. About ten years ago, Raymond Ladouceur, a sixty-five-year-old commercial Métis fisherman, started to find something new in his pickerel nets: damned ugly fish. The deformities included crooked tails, humpbacks, bulging eyes, and skin tumours. "Jesus, I was pulling them out all the time," says Ladouceur. "But we threw the deformed fish away. They weren't fit for human consumption."

In 2002, Ladouceur and other fishermen packed up two hundred pounds of the deformed fish and flew them off to Fort McMurray for study by Alberta Environment. Nobody from the government department picked up the fish over the weekend, though, and they rotted.

Like most residents of Fort Chipewyan, Ladouceur believes there is definitely something wrong with the water. He has a list of suspects. Abandoned uranium mines on the east end of the lake, for example, have been leaking for years. "God knows how much radium is in this lake," he says. Then there are the pulp mills and, of course, the tar sands and tar ponds. Ladouceur says his cousin collected yellow scum from the river downstream from the mines and dried it, and "it caught on fire." Almost everyone in Fort Chip has witnessed oil spills or leaks on the Athabasca River.

These stories caught the attention of Dr. John O'Connor. The Irishman first came to the region in 1993, working in Fort McMurray as a physician, where he earned a reputation as a straight shooter. O'Connor, who later became medical examiner for the area, frequently blasted the provincial government for the carnage on the Highway to Hell, as well as its discriminatory funding formulas for health services in Fort McMurray. In 2003, he started to administer to the health needs of the twelve hundred residents of Fort Chipewyan once a week. The diminutive fifty-one-year-old didn't know what to make of the fish tales at first, or of repeated references to the number of people "taken with cancer." But in his own work he found that the people downstream from the world's largest energy project also suffered from high rates of renal failure, lupus, and hyperthyroidism. None of his Fort McMurray patients, who rarely eat duck, moose, or fish, had these ailments.

O'Connor got especially alarmed in 2004 when he diagnosed a middle-aged man from Fort Chip with cholangiocarcinoma, a rare and painful cancer of the bile duct. O'Connor knew something about the disease because it had killed his father. "It's vicious," he says. "My father died six weeks after the diagnosis. The cancer is bad and the treatment is bad."

The cancer normally appears in only one in a hundred thousand people, and O'Connor didn't think he'd ever see it again. But he diagnosed another case of cholangiocarcinoma in Fort Chip in 2005, and yet another in 2006. It got to the point where the local toxicologist sarcastically asked him, "What are you doing to those patients up there in Fort Chip, doctor?"

Scientists don't know much about bile duct cancer, but they suspect it is caused by chemical toxins, including PAHs. Several dramatic fish studies show that white suckers, bullheads, and whitefish all tend to get bile duct cancers in waters intensely fouled by industry. After a coking plant closed its doors on Ohio's Black River, the levels of PAHs declined by 65 per cent, and liver cancers among fish dropped by 25 per cent. (The human liver works as the body's sewage system, and most toxins get exported out the bile duct.) Arsenic, another carcinogen, may also be a factor. O'Connor notes that Alberta Environment permits Suncor and Syncrude to legally dump up to 150 pounds of arsenic into the Athabasca River every year. When combined with benzene, another common tar sands toxin, arsenic dissolves a person's DNA and "leaves them open to developing cancer," O'Connor says.

O'Connor soon learned that various scientists had recommended a comprehensive health study for the community in 1996, 1999, and 2004. The province ignored them all. In the spring of 2006, O'Connor boldly asked the Alberta government for a full study of the rare cancers and other illnesses occurring in Fort Chipewyan. In response, the government produced a quick and incomplete analysis of the files of the deceased (with no peer review) that curiously excluded data from 2004 and 2005. The government concluded that cancer incidence rates in Fort Chip "were comparable to the provincial average." However, Heather Bryant, director of population health, called five cases of cholangiocarcinoma in Fort McMurray "provocative." At the same time, a Health Canada physician visited the community to prove nothing was wrong with the water. His proof consisted of grabbing a glass of water and drinking it. (Suncor and Syncrude officials carry their own bottled water when they visit Fort Chipewyan.)

O'Connor and the community of Fort Chip called the government study a sham. "Where is this cancer coming from?" the physician asked in exasperation. "Fort McMurray doesn't have this problem. I can't explain it. I'm not saying stop the oil sands, I'm just asking questions."

They were unfortunate questions to ask. Much to the horror of community elders and fishermen such as Ladouceur, members of Health Canada, Environment Canada, and Alberta Health charged O'Connor before the Alberta College of Physicians and Surgeons with causing "undue alarm" in Fort Chipewyan. The bureaucrats also accused O'Connor of overbilling, "irresponsible practices," engendering mistrust, and blocking access to medical files. It was the first time in Canada that government agencies had used a patient complaint process to silence and character-assassinate a physician. "This is not about shutting up John; this is about shutting John down," charged Michel Sauvé, the respected Fort McMurray internist.

The politically motivated investigation silenced O'Connor's inquiring voice for nearly a year. Meanwhile, the evidence for water contamination mounted. Kevin Timoney's study for the Nunee Health Authority, released in November 2007, found elevated levels of mercury, arsenic, and PAHs in local fish, water, and sediment near Fort Chip. The report raised the question of whether these contaminants were connected with the dramatic increases in fish deformities and the rare forms of cancer in the community. Timoney called for not only a major health study but an investigation of the ponds. "We have to stop building any new tailings ponds until we understand their impact on the ecosystem and the river," he said.

Dr. Jeff Short, a specialist in oil spills at the U.S. National Oceanic and Atmospheric Administration (NOAA) in Alaska, reviewed Timoney's study. Short suspects the PAHs, in particular, are coming directly from tar sands activity. He explains that PAHs can originate from either petroleum or non-petroleum sources (both leave a distinct footprint) and that the bitumen in the oil sands contains some of "the most toxic varieties."

Short describes current PAH and contaminant monitoring by RAMP, the industry and government stakeholder group, as grossly inconsistent and inadequate. "I'm quite surprised more attempts haven't been made to gauge PAHs' impact on the river," he says. "I don't get it."

Given the scale of the development in the tar sands, Short compares PAHs entering the Athabasca River from air pollution, wind erosion, and leakage from tailings ponds to a "slow-moving oil spill. What I don't know is whether it's a small or large oil spill."

Stunned by his stressful persecution and weary of the craziness in Fort McMurray, John O'Connor moved to Nova Scotia with his wife in the winter of 2007. He returns on a monthly basis as a backup doctor for Fort McKay and remains on call for the people of Fort Chipewyan. The college has absolved him of all charges except for that of causing "undue alarm."

After years of denials and delays, the Alberta Cancer Board announced in May 2008 that it would conduct a comprehensive review of cancer rates in Fort Chipewyan. Few people expect the government to find anything, although the number of cholangiocarcinoma cases in Fort Chip alone has climbed to five (three biopsied and two diagnosed). To date, no provincial or federal agency has offered to do its own review of the cancerous ponds or their seepage rates into groundwater and the river.

John O'Connor still believes the people of Fort Chip deserve a proper health study, but he doesn't think Health Canada or Alberta Health have the integrity or public trust to do one. He calls the people who live downstream from the world's largest ponds of toxic waste his "heroes." Despite overwhelming obstacles, they persevere and endure, he says. Nor has he stopped asking uncomfortable questions. "If we could reverse the flow of the river," he says, "and people in Fort McMurray had to drink the water that people in Fort Chipewyan drink, can you imagine what the reaction would be?"

SEVEN

THE FICTION OF RECLAMATION

. . . .

> "It was suggested that we should not be put off by statements that
> say we can't restore wetlands in the oil sands region...
> However, climate change in the future may make everything
> that we want to do today impossible."
>
> EXECUTIVE SUMMARY, CREATING WETLANDS IN THE
> OIL SANDS RECLAMATION WORKSHOP, CUMULATIVE ENVIRONMENTAL
> MANAGEMENT ASSOCIATION, FORT MCMURRAY, 2003

KARL CLARK, A brilliant, self-effacing chemist with Hebridean roots, spent much of his life in the boreal forest near Fort McMurray trying to figure out how to separate bitumen from sand. It took him nearly forty-five years, but the patient scientist, using his wife's washing machine, eventually perfected a hot-water process in which the bitumen frothed to the top and the sand settled to the bottom. Without Clark's fiddling and tinkering, a great many North Americans today would be walking instead of driving with gasoline made from refined bitumen. Fort McMurray named a school after him. Whenever Clark wasn't studying bitumen or fighting bureaucracy in Edmonton and Ottawa (some things never change), he camped in the bush and fished and paddled the Athabasca River. He appreciated the serenity of the forest.

In 1965, Clark got to see the first commercial application of his invention, a 45,000-barrel-a-day operation run by the Great Canadian Oil Sands Company, now Suncor. (The company opened shop officially in 1967.) The scene horrified him. Creating the mine to feed his hot-water process required the slashing and burning of thousands upon thousands of trees. The scale of the bulldozing unsettled Clark as much as the realization that he had spent most of his life making this kind of destruction possible. Before he died of cancer in December 1966, he told his daughter Mary that he could never revisit such devastation: "I don't ever want to go up again." One small open-pit mine nearly broke his heart.

Were he living today, Clark might well be in a state of perpetual grief. Much of the forest that served as his laboratory and tramping ground has been levelled. A global consortium of companies has mowed down the trees, drained the watershed, displaced the songbirds, and replaced natural wetlands with ponds full of salty waste.

The governments of Alberta and Canada, along with the multinational companies, insist not only that they'll clean up the whole mess but that rapid tar sands development is sustainable. "Alberta is proving that environmental protection and economic development can happen at the same time," promises a 2008 provincial propaganda sheet entitled "Opportunity and Balance." The Canadian Parliament, an institution less inclined to hubris, talks about groping "towards sustainable development" in its 2007 tar sands report.

Alberta's bitumen apologists swear that "work is progressing to return the disturbed land to a natural state after development, and it will be done right." The province's former ambassador to the United States, Murray Smith, even assured our number-one oil market that the industry will achieve "100 per cent long-term restoration of the lands it makes use of." Why, major tar sand companies have even planted 7.5 million tree seedlings. The Mining Association of Canada says reclaiming open-pit mines can be done with a "vision worthy of a Group of Seven artist."

According to the Alberta government, open-pit mines will eventually obliterate 1,350 square miles of forest. The government likes to minimize the scale of the destruction by saying that it's "less than 1 per cent of boreal forest area" in Canada. (In other words, it's perfectly okay to destroy small places.) Whatever the Orwellian rhetoric, the forest-top removal will cover an area four times larger than that of New York City. Classical scholars and students of collapsed civilizations should note that this fossil-fuel excavation site will be three times greater than that occupied by the ancient city of Angkor Wat in Cambodia. Outdoor enthusiasts can imagine half of Banff National Park flattened and excavated.

Even at that, the mines make up only a small part of the wreckage created by the megaproject. The Alberta government has leased an additional 23,000 square miles of land (and another 30,000 square miles await global investors) for in situ projects, including steam-assisted gravity drainage. The Canadian Rocky Mountain Parks, which encompass 9,000 square miles and include Jasper, Banff, Yoho and Kootenay, could fit into this planned industrial zone about six times. As noted, SAGD development will slice and dice the land with thousands of industrial well sites, seismic lines, pipelines, and roads. This fragmentation will transform the forest into a bitumen park, exterminating the population of woodland caribou and decimating songbirds home from their winter in the tropics. Seismic lines, which make a forest look like an engineered spiderweb, typically need more than one hundred years to fill in with trees again. Yet the government has no tight guidelines for reclaiming forest ruined by SAGD.

Government definitions of reclamation exhibit a genuine vagueness as well as a preference for mechanics over biology. According to Alberta's Environmental Protection and Enhancement Act, reclamation is mostly about "stabilization, contouring, maintenance, conditioning or reconstruction of the surface of the land." Operators of the open-pit mines must "conserve and reclaim disturbed land to an equivalent land capability." Doing so will earn them a certificate proving the deed done. Industry-friendly scientists talk about creating "a self-sustaining ecosystem with no long-term toxicity." Those reassured by such academic language might want to consider the actual pace of reclamation: after nearly fifty years of mining, the provincial government has certified only 257 acres of forest, or 0.2 per cent of the land dug up since 1963. Even industry admits that reclamation has moved more slowly than bitumen in a pipeline.

At the Wood Bison Viewpoint, about twenty miles north of Fort McMurray, Syncrude advertises the future of reclamation with what *Oil and Gas Weekly* calls "a scene reminiscent of the way things once were: human, land and beast coexisting in harmonious simplicity." Syncrude opened the reassuring public site in 1995, right next to an "emergency meeting point" to be used in the event of a toxic spill or an upgrader fire. It's not really a forest but a fenced grassland that supports some three hundred wood bison. Syncrude says their bison win awards at country fairs in Saskatchewan, though it's unclear if anyone eats the meat. Chemically corroded signs at the viewpoint promise that reclamation will turn moonscapes into "areas ideal for hiking, boating, fishing and wildlife viewing" within twenty years. One sign even shows a happy couple sailing across a reclaimed tailings pond. Syncrude, a leader in land reclamation, spent one-fifth of 1 per cent of its budget on reclamation in 2005.

As far back as 1977, a group of soil specialists at the University of Alberta reported to Syncrude that any effort to regrow plants and trees on the mined area would have to overcome "salinity, oil, low fertility, erosion and unfavourable soil reaction." Ronald Pauls, a forest reclaimer with Syncrude, echoed these sentiments at the Twelfth Vertebrate Pest Conference in 1986, confessing that planting greenhouse-pampered spruce and jack pine in the tar sands waste was problematic because of "low seedling survival and slow growth rates." Pauls also blamed hungry meadow voles and deer mice for slowing down reclamation efforts by munching on seedlings.

In 2002, in the *Journal of Environmental Quality,* Syncrude investigators again reported on the shortage of solutions. Soil peppered with mining waste typically contains so much salt that young seedlings of jack pine, a tree native to the region, "exhibited growth reduction and visible signs of injury." Barley, another candidate for early reclamation, didn't fare much better. Studies have found that red-osier dogwood and hybrid aspen can accommodate salts more easily.

But finding salt-tolerant trees and shrubbery isn't the only obstacle to reclamation. Wetlands, the kidneys of any landscape, purify water, control erosion, prevent flooding, moderate climate change, and serve as homes for birds and fish. Although boreal wetlands once covered half of the region dug up by the mines, the Alberta government has yet to offer any criteria for their reclamation. In truth, no one really knows how to do it. At the conclusion of a 2003 symposium on wetland restoration in the tar sands, a group of well-known scientists wrote that the mining boom was being allowed to proceed "with little real knowledge . . . of how [the wetlands] will be reclaimed." Far too little money was going into reclamation solutions, said the scientists, and "the regulatory agency has no apparent commitment to really recreate wetlands on the landscape." Worst of all, they doubted that there would be enough water at the end of the day to restore anything.

Alberta's Guideline for Wetland Establishment on Reclaimed Oil Sands Leases, revised in 2007, still offers no criteria. Slipped in among its many platitudes is the admission that an "atmosphere of uncertainty" still hangs over wetland reclamation. Much of this uncertainty, which the government calls "knowledge gaps," directly relates to climate change. Global warming, which rapid tar sands development has duly stoked, could undermine many reclamation plots by drying them up. Climate change has already raised temperatures in the region by three to five degrees Fahrenheit and will soon turn up the thermostat another three to five degrees more. This remarkable warming will reduce rainfall and accelerate evaporation. Higher temperatures will concentrate fish-killing salts and naphthenic acids in remaining water bodies. Nobody knows how groundwater will flow below the engineered and dried-out landscape, either. According to the 2007 guideline, it is "unclear whether natural fens and bogs will persist in the oil sands region." After twenty years of research, the report says "reclamation of wetlands on oil sands leases is still in its infancy."

Reclamation science hasn't advanced much since Syncrude's realistic 1977 report. Yet company after company continues to file cheerful Environment Impact Assessments. The documents aim to assure unscientific minds that industry has figured out how to grow a boreal forest from scratch. Imperial's Kearl project, for example, will disturb

57,000 acres of forest and wetlands by digging four open-pit mines. Yet by 2060, the company says, the reclaimed landscape will "replicate the stability and robustness of the original natural systems." This novel transformation will include six end-pit lakes filled with waste water and toxic tailings as well as "a dynamic drainage system that accommodates evolutionary changes but does not accelerate erosion or cause unacceptable environmental effects." Blueberries and balsam poplar, "a salt tolerant tree species," will dominate this new world. Imperial, a serious supporter of attacks on climate change science, does admit that global warming, combined with salinity issues, might affect "some soil and vegetation targets."

Syncrude vowed in a 2004 paper on reclamation "to return the land we disturb to a stable, biologically self-sustaining state. This means creating a landscape that has a productive capability equal to, if not better than, its condition before mining began." The application for CNRL's Horizon Project, an open-pit mine affecting 115,000 acres, assured regulators in 2002 that the company would have no problems remaking a forest, or what it calls "a trafficable" landscape: "Mitigation paired with reclamation assumes a post project success rate of 100 per cent. Uncertainty about reclamation methods is assumed to be resolved with ongoing reclamation monitoring and research." Chris Jones, chief operating officer of Albian Sands Energy, told a public committee in 2006 that his firm's ultimate reclamation goal was "to achieve maintenance-free, self sustaining ecosystems with a capability that is equivalent to predevelopment conditions." Almost every multinational now asserts that something called "adaptive management" will rescue reclamation from "the atmosphere of uncertainity." Adaptive management means learning by doing, even when you don't know what you are doing. It's also what retreating soldiers do when their mission fails.

Dr. Lee Foote, a wetland specialist at the University of Alberta, conservatively calculates that it will cost at least $10,000 to reclaim one acre of lost wetland in the tar sands. In jurisdictions such as West Virginia, industry must restore three acres of wetland for every acre lost because wetlands perform so many crucial ecological services. Some scientists suggest a ratio of 10 to 1 would be healthier. Given that 237,000 acres of wetlands have been dug up to date in the tar sands, Foote estimates that the eventual cost of replacing wetlands could range anywhere from $2.4 billion to $24 billion. He qualifies even those figures. "It's a significant liability if it can be done at all," says Foote.

Many of the companies digging up wetlands along the Athabasca River, such as Exxon (part of the Syncrude consortium) and Shell, have already left an expensive legacy in Louisiana. Like Alberta, the bayou state has been a petrostate for years, producing 30 per cent of the domestic crude oil in the United States. For more than three decades, the state's oil industry compromised coastal marshes and wetlands with ten thousand miles of navigational canals and thirty-five thousand miles of pipelines. These industrial channels, carved into swamps, invited salt water inland, which in turn killed the trees and grasses that kept the marshes intact. The U.S. Geological Survey suspects that the sucking of oil from the ground has also abetted the erosion. Since the 1930s, nearly one-fifth of the state's precious delta has disappeared into the Gulf of Mexico. In fact, the loss of coastal wetlands now threatens the security of the industry that helped to destroy them. Without the protective buffer of wetlands, wells, pipelines, refineries, and platforms are more vulnerable to storms and hurricanes. Even *Fortune* magazine publishes stories about "how the energy business is drowning Louisiana." Federal scientists now lament that the state loses a wetland the size of a football field every thirty-eight minutes.

Young Canadians sensibly wonder if future generations will have the gall to call the destruction of boreal forest and bogs for bitumen production "sustainable development." In 2004, a University of Alberta business student concluded his study of Alberta's reclamation record by asking if future governments wouldn't identify the tar sands "as an abandoned environmental catastrophe whose burden of reclamation and remediation will be borne by Canadian citizens." In 2006, another university student, the son of a tar sands manager, offered an equally stark assessment. He noted that no firm had ever closed a bitumen mine and wondered if environmental ideals wouldn't eventually take a backseat to consumer demands for oil. "Will future Mars-mission trainees spend their time in the barren and lifeless remains of the abandoned oil sands mines?" he asked.

Miners offer anonymous Internet comments along the same lines. "I figure we should be restricting open pit mining to a limited area (only those already started) and only allow new areas to be the size of reclaimed areas," suggested one Fort McMurray worker in a talk-back forum after the airing of the CBC documentary *The Tar Sands: The Selling of Alberta*. "By reclaimed I mean self-sustaining natural forest (or desert if the climate changes). This way we can continue open pit mining and be sure that when the music stops the size of the devastation will not be any larger than it is now. You know they (the companies) will not clean up after themselves unless we make them and we can't do that after the companies leave and the money is gone."

CANADIANS, A LONGTIME mining people, have a nasty habit of leaving behind Mars-mission landscapes for taxpayers to clean up. Over the last thirty years, federal regulators have repeatedly allowed mining companies to walk away from their industrial waste piles. Today, ten thousand abandoned mines litter the country. The public will either have to clean up these toxic legacies or learn to enjoy water spiced with arsenic and cyanide. No responsible Canadian authority, it seems, bothered with security deposits from the companies.

Canadian taxpayers, who made $150 million in royalties from mining activities between 1966 and 2002, have spent more than $4 billion tidying up scores of contaminated sites, including 233,000 tons of arsenic waste at the Giant Mine in the Northwest Territories and the seepage of acids and heavy metals from the Britannia Mine in British Columbia, which until recently was one of the largest metal-pollution sources in North America.

In 2002, Canada's auditor general, Sheila Fraser, declared the sorry state of northern mine abandonment "far from a good example of environmental excellence." She found "an urgent problem" at the Northwest Territories' Colomac Mine site, which at the time produced enough polluted water to flood ninety-three football fields. There was enough arsenic dust at the Giant Mine in Yellowknife to fill an eleven-storey building. Her report documented heavy contamination of surface water with zinc at the Yukon's Faro Mine. It found "an environmental mess" at the Yukon's Mount Nansen Mine, along with a record of corporate water-law breaking. Taxpayers paid nearly half a million dollars a year in fuel bills to run the equipment to treat the mine's toxic tailings water. The auditor general concluded that government's "band-aid approach" to containing waste from abandoned mines was "not sustainable in the long term." (Since 2002, industry has actively lobbied to weaken rules on mine closures in the Yukon and British Columbia.)

When the auditor general revisited the issue in 2008, she found that the federal government had made some progress, spending hundreds of millions of taxpayer dollars on the orphaned mines that now make up 30 per cent of Canada's contaminated sites. But there are still big messes. Water at the Nansen Mine will require treatment for several hundred years more, and the arsenic cleanup at Giant Mine will threaten groundwater well beyond 2020. Moreover, the auditor general said, it was not clear how cleanups would help the government to "eliminate the financial liability of known contaminated sites."

Reclamation liability in the tar sands hasn't made any national headlines yet. The Alberta government claims to hold nearly half a billion dollars in security deposits in case tar sands miners go broke, get tired, or simply leave the scene. That amounts to nearly $5,000 per acre, yet the only patch of forest certified so far as reclaimed (and it had no tailings pond) cost approximately $46,000 per acre. It appears that the designers of Alberta's security program failed basic math in high school. In February 2007, the Alberta government admitted that, after fifty years of mining, it was still developing a Mine Liability Management Program. The current security program doesn't apply to decommissioning billions of dollars' worth of upgraders, pipelines, coke ovens, and the like. A 2008 study on reclamation by the Pembina Institute concluded that Alberta's tar sands security program lacks transparency and that information about costs, bonds, and validation of reclamation "are not publicly available or readily accessible."

The province's reclamation record for abandoned oil and gas well sites and facilities is scandalous. The system is largely self-regulating, and the standard for restoring land disturbed by well pads has dropped to 60 per cent of original soil content. Unlike the governments of Wyoming or Alaska, Alberta allows companies to drill wells without posting any serious reclamation bond for a cleanup. (Alaska charges $100,000 an oil well; Alberta charges a $10,000 licensing fee and then lets companies drill as many wells as they want.) In the last ten years, the number of abandoned sites and facilities has risen faster than the price of gasoline. More than a hundred thousand abandoned sites now pose a threat to groundwater and agriculture throughout rural Alberta. The government's own records show that it has knowingly permitted the province's reclamation liability to rocket from $6 billion in 2003 to $18 billion in 2008. If not addressed, the public cost of cleanup could eventually consume more than two decades' worth of royalties from the tar sands. The ERCB holds but $35 million in security deposits for $18-billion worth of abandoned oil field detritus.

If history, as the Greek storyteller Thucydides maintained, "is philosophy learned from examples," then the Alberta and Canadian governments are profoundly unphilosophical. The uncomfortable truth remains simply this: the rapid mining of the boreal forest has outpaced the science on the reclamation of wetlands, soil, and forest uplands by decades. No one has a handle on the real costs of reclamation. Security deposits remain laughably inadequate. And both Alberta and Canada have an appalling record of environmental negligence and disregard for taxpayers.

Reclamation in the tar sands now amounts to little more than putting lipstick on a corpse. Unless Alberta and Canada soon address the pace, effectiveness, and transparency of reclamation, a rich forest will become an impoverished industrial park littered with salts, grass, polluted water, and spindly trees. It might, with a bit of luck and some regular rainfall, eventually resemble a third-rate golf course in the Sudan.

EIGHT

DRAGONS AND PIPELINES

. . . .

"The 'dirty-oil' thing is unfair...there is an environmental
footprint associated with all forms of oil."
ROBERT JONES, KEYSTONE PIPELINE CEO, EDMONTON, 2008

WHEN DR. DONALD R. BLAKE, one of the world's foremost experts at
measuring air pollution, first set eyes on Upgrader Alley, Alberta's
industrial heartland just east of Edmonton, the scale of the enterprise
stunned him. "Wow," said Blake, a scientist from the University of California, Irvine. "Sacrificing all this crop land just to look as industrialized
as California. I guess wherever you find oil and gas folks, you find the
same intellect, which is you go and do what you've got to do."

This brand of intellect has already made Upgrader Alley the largest
petrochemical complex in Canada and the second largest in North
America. Rapid tar sands development may shoot it into first place. By
2020, three provincial pipelines from Fort McMurray will ferry three
million barrels of raw bitumen a day to Upgrader Alley, and in so doing
transform the counties of Strathcona, Sturgeon, and Lamont and the
City of Fort Saskatchewan into a "world class energy hub." Just about
every company with a mine or SAGD project in Fort McMurray, from
Total to Statoil, has joined the rush to build nearly $45 billion worth of

upgraders, refineries, and gasification plants. The colossal development will not only industrialize a 180-square-mile piece of prime farmland straddling the North Saskatchewan River (an area half the size of Edmonton) but consume the same amount of water as one million Edmontonians. The situation is another example of Alberta gone wild.

As do most practical Americans, Blake appreciates the geopolitics bitumenizing the region. "I don't see any way around it," he says. "You guys have all these fantastic resources and can turn them into fuel. The United States would be doing the same thing." But he does find the scale of the industrialization "scary." The region already has some of the most polluted air on the planet. Blake's lab is ranked number one out of thirty facilities in the world for identifying and quantifying the trace gases that make smog and ozone. In 2005, he collected sixty-four whole-air samples near the building site for BA Energy's upgrader.

The results dumbfounded him. In so-called clean Canada, Blake found levels of pollution three to four times higher than he had recorded in oil-rich Texas and seven to ten times higher than in the U.S. Midwest. "I am surprised that trace gas concentrations in a rural location in Canada are in many cases considerably more enhanced than in a polluted urban center like New York City...and many cities in China," says Blake. Emissions from the oil and gas industry, he concluded, have had "a surprisingly large impact on air quality in the area."

But where Blake found remarkable volumes of dirty air, government and industry can smell only opportunity. According to their enthusiastic brochures and PowerPoint presentations, the heartland is "the world's most attractive location for petrochemical investment today." A landscape that once supported potato and dairy farms will soon be dotted with supersized industrial bitumen factories exporting synthetic crude and jet fuel to Asia and the United States. "Successful upgrading to finished product could add billions to the Alberta and Canadian economy and broaden Alberta's markets for value-added products," predicted Houston analyst and chemical engineer David Netzer. According to Netzer, plunking as many as fifteen upgraders in one spot will not only reduce their environmental footprint but foster "operational synergies and lower costs." The whole enterprise will supposedly be sustainable too, although industry admits that "air quality and climate change issues have yet to be resolved."

Bitumen upgraders are among the world's most proficient air polluters because, as the 2006 *Alberta's Heavy Oil and Oil Sands* guidebook notes, they are "all about turning a sow's ear into a silk purse." Removing impurities from bitumen or adding hydrogen requires dramatic feats of engineering that produce two to three times more nitrogen dioxides (a smog maker), sulfur dioxide (an acid-rain promoter), volatile organic compounds (an ozone developer), and particulate matter (a lung and heart killer) than the refining of conventional oil. In 2005, Syncrude spewed out 219,054,364 pounds of toxic air pollutants, making it Canada's fourth-largest air fouler. According to the Pembina Institute, Fort McMurray's four upgraders release nearly three hundred tons of sulfur into the air every day (and acid rain is falling again in Saskatchewan). An upgrader's byproducts, as a 2006 oil sands presentation noted in Houston, "are plentiful and nasty: sulphur and coke." The scale of the cracking, coking, and heating is so immense and complex that upgraders often behave like temperamental dragons. They routinely catch fire—Suncor had crippling fires in 2005 and 2007—and leak like hell, even with proper maintenance.

Intense tar sands development combined with lots of upgrading blackens the air. A 2007 Alberta Environment report found that concentrations of hydrogen sulfide had increased in the oil sands by 30 per cent to 175 per cent since 1999; that nitrogen oxide increased by 23 per cent; and that the highest provincial concentrations of particulate matter occurred at Suncor's Millennium Mine. Just about every immune-busting and lung-clogging pollutant will double or triple in volume in the Fort McMurray region by 2020.

But from a government's point of view, a multibillion-dollar upgrader is much more appealing than a farm. A typical midsized upgrader, for example, can pipe $450 million worth of taxes into federal and provincial coffers every year for twenty-five years. The construction of half a dozen upgraders can employ twenty thousand people for a decade and keep the economy growing like an algae bloom. According to the economic development department at Strathcona County, the construction of each new upgrader will use enough cable to stretch from Vancouver to Toronto; enough concrete to build a sidewalk from Edmonton to Fort McMurray; and enough steel to build a railroad track from Toronto to Ottawa. Other studies note that the demand for supplies, services, and maintenance could exceed $100 billion over fifty years.

The advantages of upgrading bitumen in Canada, as opposed to shipping the resource south, are considerable. In 2002, heavy oil expert Maurice Dusseault calculated in a report for Alberta Energy that the export of 300,000 barrels of heavy crude or bitumen to Chicago, Minneapolis, Kansas City, and Billings, Montana, represented an annual loss of $1 billion. Relative to conventional crude, bitumen typically sells at such a heavy discount that U.S. refineries equipped to handle the product can turn over incredible profits. "The lost profits and lost opportunities are simply too large to ignore," concluded Dusseault. But the Alberta government did ignore them, and by 2007 bitumen's lower price differential amounted to a loss of $2 billion a year. Money is lost whenever raw bitumen is exported.

Although the economic argument for upgrading bitumen in Alberta or Canada is strong, the scale and pace of the upgrader boom has become a nightmare for ordinary citizens. Consider, for example, the approval of the North West Upgrader, an independent merchant operation that will refine up to 200,000 barrels of bitumen a day.

In 2007, ERCB blessed the $4-billion project after a brief hearing. Local landowners, many of whose families have lived in the region for three generations, opposed the project because of the province's Gillette Syndrome: explosive growth in traffic, crime, noise, and air pollution, along with questionable emergency response plans. Residents despaired over local politicians who swore that they wouldn't live any closer than a mile to an upgrader but then approved minimum setbacks of five hundred yards for single-family homes. One family noted they felt like "a sacrificial lamb whose lives are being determined by people who do not even live within the area." Rural residents also defended their conservative belief, in the libertarian land of bitumen, that "planning should come before development," which "is simply not happening here."

Wayne Groot, a grower of high-quality seed potatoes (a global famine fighter), told the remarkably oblivious ERCB panel that the destruction of fifty thousand acres of sandy loam soils was unforgivable; he had once dreamed of passing on a farm to his children, not "a heavy industrial war zone." There was a huge difference "between us moving next to an Upgrader versus an Upgrader moving next to us," said Groot. Then the potato farmer eloquently spelled out the unreality of the tar sands tsunami:

> When I look around our province and see what is happening to it, I cringe. I see a rampant rush to exploit our unrenewable natural resources for economic gain, with negligible thought put into any of the consequences to our natural or socioeconomic environment. Our economy is out of control and we are all caught up in the race for the biggest home and the most toys, yet I think we all know that this is not sustainable and will come crashing down. And when it does, most, if not all, of us will suffer.
>
> We talk about the Alberta Advantage, yet I still have no idea what that really means. I don't think it means much to anyone who is trying to buy a new home. I don't think it means anything to small-business owners who cannot find anyone to work for them for a modest wage... I hope that twenty years from now, I can still call this a beautiful province. At present, I have my doubts.

Landowners at the hearing also raised questions about the integrity of a buyout program. In 1999, the ERCB had agreed that the expansion of Upgrader Alley was "ultimately not acceptable without relocation of the residents of the area and in future the Board may not be able to approve additional projects in the area until the issue has been resolved." But the board did nothing. Industry eventually set up a Voluntary Purchase Plan Program (VPPP) for those local residents who didn't want to breathe sulfur dioxide and hydrogen sulfide or listen to alarm bells all night.

Landowners quickly dubbed the VPPP a "victim persecution policy." Rancher Laura Martin outlined the program's unique deficiencies one by one. She and her husband got so fed up with sick animals, oil spills, and traffic accidents caused by Upgrader Alley that they moved to Saskatchewan in the spring of 2007. Martin noted that the VPPP was run by industry and that industry promoters chose who was affected adversely enough to qualify. In 2006, the VPPP, with a measly budget of $3 million, bought out only three of twenty-four applicants. The province refused to participate in the program or help Albertans displaced by the upgrader boom.

Landowners accepted into the program also found their civil rights truncated: they couldn't file an objection to a megaproject in their backyard without being "put on hold," which meant losing their spot in the program. Lastly, the VPPP required landowners to sign a confidentiality clause that forbade them from commenting about their experiences or testifying at public hearings. The program contained so many loopholes that Martin wondered if Albertans would need "respirators and oxygen" before they qualified for compensation.

Donald Blake, an air pollution expert, testified at the hearing on behalf of the local farmers: the Northeast Sturgeon County Industrial Landowners. At their request he again sampled air quality in Upgrader Alley, this time at the site proposed for the North West Upgrader. In 2006, he took seventy whole-air samples and analyzed them for eighty-one compounds, including methane, carbon monoxide, propane, and benzene. This time he found pollution levels five times higher than those recorded in Texas. The pollution also surpassed levels found in forty-three of China's most polluted cities. Near Shell's Scotford Refinery, Blake found concentrations of ethane and benzene fifty times greater than elsewhere in the region, indicating massive leaks as well as a health hazard to refinery workers. He also found ozone levels high enough to stunt the growth of local agricultural crops downwind.

Blake's detailed report directly contradicted the nothing-to-worry-about results of air monitoring done by the Fort Air Partnership (FAP), an industry group, and Environment Canada. He explained that the monitoring stations of those two bodies were conveniently located three to six miles downwind, where "you will always find low values." Blake recommended a more rigorous air sampling program and more monitoring sites because "the air in the study area is already markedly polluted." Local residents have long complained that the FAP has failed by design or neglect to install active air monitoring stations within the boundaries of the most-affected communities.

None of these revelations moved the ERCB or even delayed the project. In the end, the board ruled that the bitumen money factory was in the public interest, partly because "no concerns were expressed by any participants with respect to the need for the North West upgrader project." The board acknowledged that the VPPP "seemed to continuously change and it was not a transparent program," yet the project was approved without a fair buyout program in place for displaced rural residents. Without a shred of embarrassment, the board also admitted that "no comprehensive regional monitoring program assessing the effects of regional emissions on terrestrial ecosystems and potential soil acidification" now existed and that pollution already exceeded Alberta and federal guidelines on site.

The ERCB ultimately recommended that Alberta Environment work on the issue. Two upgraders are now being constructed in Upgrader Alley, and another ten to fifteen bitumen refineries are in the planning stages. Although Alberta Environment says it will cap nitrogen oxide pollution levels at 25,000 tons a year for the region, it states clearly that it wants to "optimize growth." The big challenge, the department says, is not cancerous air or unhealthy citizens but the fact that "the scale of the opportunity exceeds the capacity of the current management system." Given these businesslike sentiments, no one honestly expects the air to get cleaner. If the ERCB's 2007 public record on the regulation of natural gas–processing refineries is any indication of how bitumen upgraders will audited for pollution, then rural Albertans can expect compliance rates of 52 per cent.

In 2008, Donald Blake returned to Alberta to testify at a public hearing on the $17-billion PetroCanada Sturgeon Upgrader about air quality in Upgrader Alley. This time, Blake reported finding the highest levels of methane, a global warmer, ever recorded in a northern latitude. Just four hundred yards downwind of the Shell Scotford Upgrader, Blake had also detected styrene levels four times higher than those measured in Mexico City, one of the most polluted cities in the world. "It is remarkable that these gases in rural Alberta are already on the same scale as those in Mexico City," he said in his report. Blake discovered at the hearing that the industry-funded air monitoring program, Fort Air Partnership, was staffed by volunteers, that the program routinely failed to report pollution violations, and that it had operated eight stations for two years with equipment rated as dirty or malfunctioning by two successive Alberta Environment audits. Such third-rate efforts in one of the most industrialized air sheds in Canada, said Blake later, "reeked of coverup and sleight of hand."

OUT-OF-CONTROL AIR POLLUTION and the abuse of property rights and civil rights aren't the only issues raised by the expansion of Upgrader Alley. Upgraders, like the processes that supply them with bitumen, gulp lakes of water. The North West Upgrader, for example, will annually use up to 1.2 billion gallons of water from the North Saskatchewan, a river only a third the size of the Athabasca. Most of that water will be used for cooling or dumping waste energy.

In 2007, a report done by the engineering firm Morrison Hershfield for Strathcona and Sturgeon counties added up the water footprint for the upgrader boom. Each new facility will require anywhere between 3.5 and 4.5 million gallons of water a day, the equivalent of six to eight Olympic-sized swimming pools. By 2026, their collective daily thirst could amount to ten times as much. In contrast, the city of Edmonton uses 77 million gallons a day and returns most of that water to the river in treated form. The upgraders won't do that: some 70 per cent of the water will be consumed or lost to evaporation.

The oil patch is the second-highest water user in the North Saskatchewan River basin (using 18 per cent of water withdrawals). The upgrader boom will make the petroleum sector number one. A 2007 report for the North Saskatchewan Watershed Alliance says that "nearly all of the projected increase in surface water use will be in the petroleum sector." By 2015, the upgraders' demands on river water will increase by 278 per cent; by 2025, 339 per cent. John Thompson, author of the report, says the absence of an authoritative study on the river's ecosystem, an Alberta trademark, leaves a big hole. "We don't know what it takes to maintain the river's health."

Providing energy for the upgraders will also take a toll on water. Sherritt International and its investment partner, the Ontario Teachers' Pension Plan, are proposing to strip-mine a 120-square-mile area just east of Upgrader Alley for coal. A gasification plant would render the coal into synthetic gas and hydrogen to help power the upgraders. Current estimates suggest that the project will consume somewhere between 70 million and 317 million cubic feet of water from the North Saskatchewan annually. Strip-mining farmland will also "affect groundwater aquifers and surface water hydrology."

In December 2007, Alberta Environment released a new framework for the river that concluded "ample capacity exists in the North Saskatchewan River to support a healthy industry and growing population." The report okayed all current projects but noted that "the current level of proposed development calls out for a comprehensive review." The department also noted that "water quality could continue to decline without cumulative limits in place and actions to mitigate further impacts."

The technological innovation championed by both industry and government to address potential water shortages involves using Edmonton's grey water. Instead of allowing a dozen upgraders to stick individual straws in the North Saskatchewan, the city's utility company,

EPCOR, would pipe the city's treated wastewater to the upgraders. This would both lessen the load of chemicals on the river (treated wastewater contains nitrogen and phosphorus) and provide a secure supply for industry. "It's a good solution for the industrial heartland and the river," says Joe Gysel, EPCOR's vice-president of marketing and business development.

But even if they use wastewater—a common practice in water-short California and Colorado—the upgraders will continue to drain the river. Famed water ecologist David Schindler calls the framework's claims of ample capacity "pretty hollow." He notes that the framework avoids any mention of declining river flows, disappearing wetlands, or the expected effects of climate change. The largely industry-dominated panel that drew up the framework, he adds, was "completely one-sided and way beyond its depth."

Historically, the North Saskatchewan River has been subject to extreme variations in flow, says Dave Sauchyn, a climate change specialist at the University of Regina. In 1796, a drought year, the Hudson's Bay Company had trouble moving furs, "there being no water in the river," as an eyewitness put it. Sauchyn says that eighty years of record-keeping on the river are insufficient to predict variability in water availability. He adds that both the lowest and one of the highest flows recorded on the river took place between 2001 and 2005. Sauchyn, who has recently begun to study the impact of climate change on the river, already has a "gut reaction" to the idea of putting as many as fifteen upgraders along its banks: "They should be thinking about whether it's judicious to proceed, or how to store water during low flows."

WHILE RAPID DEVELOPMENT of the tar sands turns Alberta's industrial heartland into another dysfunctional Fort McMurray, North American pipeline companies are busy constructing $31 billion worth of bitumen highways. In particular, the National Energy Board has approved three massive pipelines that will help to move 1.5 million barrels of raw bitumen south of the border for upgrading at U.S. refineries. (That volume currently equals total production in the tar sands.) The $3-billion

Alberta Clipper Project, "the largest expansion in Enbridge's history," will take bitumen to Superior, Wisconsin, while TransCanada's $5-billion Keystone Pipeline will take bitumen from Hardisty, Alberta, across seven states to Cushing, Oklahoma. A proposed $7-billion expansion could also take the product to refineries on the U.S. Gulf Coast. Enbridge and ExxonMobil are proposing to ferry more bitumen to the Gulf via the Texas Access line from Patoka, Illinois. And Enbridge's Southern Lights Project will transport the diluent (naphtha, condensate, or light oil) to make all these exports possible. As one analysis explains, the Southern Lights pipeline will be "an enormous bitumen exporting conveyor loop" helping the United States to motor more raw bitumen south.

Enbridge has plans to build another $5-billion global superhighway, the Northern Gateway Project. It would take 525,000 barrels of bitumen a day from Edmonton's Upgrader Alley to the deep-water port of Kitimat, B.C., for sale in California and Asian markets. The dual 700-mile-long pipeline would also import 200,000 barrels of condensate or diluent from Russia to help lubricate the export line. Enbridge calls the Gateway Project "an important part of Canada's energy future." But the scheme would bring as many as thirteen supertankers a month to Kitimat and expose B.C. coastal waters to average spills of one thousand barrels every four years and ten thousand barrels every nine years.

Superhighways for bitumen have enormous implications for Canada. The pipelines will determine the nation's economic future by accelerating the pace of tar sands exploitation and liquidation. They will also return Canada to its roots as a provider of raw, undervalued staples. According to an economic report by Informetrica, the export of 400,000 barrels per day represents the loss of eighteen thousand jobs and a 0.2 per cent subtraction from Canada's GDP. The export of 1.5 million barrels a day adds up to the wholesale extradition of nearly sixty thousand jobs and billions of dollars. Under the rules of the North American Free Trade Agreement, Canada will not be able to claw back these exports and will permanently lose the economic benefits of processing the resource at home. In other words, the pipelines will probably knock off a million barrels a day or more of planned upgrader capacity in the heartland.

Not only that: the rapid export scheme approved by the National Energy Board will decrease the proportion of bitumen serving the Canadian market from 36 per cent to 29 per cent. The proportion going south will climb from 64 per cent to 70 per cent. Given that NAFTA rules force Canada to maintain a proportional export to the United States (Mexico wisely rejected the proportionality clause on energy exports), these three new pipelines will undermine our nation's energy security. In the event of an international energy emergency, the pipelines guarantee that the United States will get the greatest share of Canadian oil. "It hasn't dawned on most Canadians that their government has signed away their right to have first access to their own energy supplies," says Gordon Laxer, director of the Parkland Institute.

A 2008 petition to the federal government by the Communications, Energy & Paperworkers Union of Canada (CEP) argued that the bitumen superhighways will leave most of Canada "vulnerable to offshore supply disruptions" and will undermine "the potential to establish a diversified and sustainable oil and gas industry for Canada." CEP added that the National Energy Board "has entirely lost sight of its statutory mandate and has all but abandoned its critical role as the guardian of the public interest." (One of the key goals of the 1996 *Declaration of Opportunity* was to make Canada self-sufficient in oil.)

The export of bitumen to retrofitted U.S. refineries will dirty waterways, air sheds, and local communities. About 70 per cent of current refinery expansion proposed in the United States (a total of seventeen renovations and five new refineries) is dedicated to bitumen from the tar sands. Companies such as BP, Marathon, Shell, and ConocoPhillips have announced plans to expand and refit nearly half a dozen older refineries in the Great Lakes region to process bitumen. The people who live along the largest freshwater system on Earth now fear they will be breathing air as bad as that in Edmonton's Upgrader Alley.

Last year the government of Indiana approved a $3.8-billion expansion at BP Whiting refinery on Lake Michigan. The state's permit allowed BP to increase its ammonia wastes by half and its industrial wastes by one-third. A political storm erupted as soon as citizens learned about the Albertalike exemptions. When the mayor of Chicago protested the dumping of refinery waste into Lake Michigan as "unacceptable," and Republican congressman Fred Upton called it "wholly unacceptable," BP changed its plans. The refinery will also emit fantastic volumes of carbon dioxide into the atmosphere: every year, the equivalent of 340,000 vehicles.

In Wisconsin, Murphy Oil proposes a $6-billion expansion at its refinery on Lake Superior to enable it to refine bitumen too. The project will consume five million gallons of water per day from the lake, boost the refinery's energy demand twelvefold, and destroy four hundred acres of wetlands—what one environmentalist called "the largest wetlands filling in Wisconsin since the passage of the U.S. Clean Water Act of 1972." Marathon Oil hopes to add a $1.5-billion expansion to its refinery in Detroit, a place already ranked the ninth worst in the United States for short-term particle pollution.

On the Canadian side of the Great Lakes, refineries are expanding in Sarnia's notorious Chemical Valley. The area already boasts more than sixty-five petrochemical facilities, including a Suncor refinery that has been upgrading bitumen for fifty-five years. Shell wants to add a bitumen upgrader to the mix, and Suncor just completed a billion-dollar addition to handle more dirty oil.

The region currently suffers from some of the worst air pollution in Canada. Industrial waste from Chemical Valley has feminized male snapping turtles in the St. Clair River, turned 45 per cent of the whitefish in Lake St. Clair "intersexual," and exposed two thousand members of the Aamjiwnaang First Nation to a daily cocktail of 105 carcinogens and gender-benders. Newborn girls outnumber boys by two to one on the reserve. Two-thirds of the children have asthma, and 40 per cent of pregnant women experience miscarriages. Calls for a thorough federal investigation have gone unheeded. Environment Canada has never bothered to do a cumulative impact study, and it's unlikely any responsible authority ever will.

A mother born and raised in Chemical Valley recently posted her thoughts on MyMcMurray.com about the national and personal compromises forged by the inexorable expansion of the tar sands:

> I have seen my fair share of pollution, illness, etc. There are forms of cancer here that do not exist elsewhere but I was raised on that money. My father, although he never liked his job, has worked in the plants for 28 years. His job paid the bills, clothes, food etc. No one around here likes what the plants are doing, but for the sake of our economy very few speak out. I am sure it is the same in Fort Mac. This will not resolve the issues, but finding a solution is going to take diplomacy and the right people to make the right choices. I lost my step father and father in law to rare cancers—they both worked in the plants. I have seen and heard so many horror stories about what these plants are doing to us. But when you have a family to support, what do you do? You do what you have to, even if you don't like it. This is not to say that people should not pursue better, but that most are afraid to do so.

The marketplace and quislinglike regulators are directing our country's insecure economic future without a vote or even so much as a polite conversation over coffee. Canadians can now choose between two nightmares: an air-fouling, river-drinking economy that upgrades the world's dirtiest hydrocarbon on prime farmland or a traditional staples economy that exports cheap bitumen and thousands of jobs to polluting refineries in China, the Gulf Coast, and the Great Lakes while making Eastern Canada ever more dependent on the uncertain supply of foreign oil. There is currently no plan C.

NINE

CARBON: A WEDDING AND A FUNERAL

....

"If you are investing in tar sands or shale oil, then you
have a portfolio that is crammed with sub-prime carbon assets."
AL GORE, FORMER U.S. VICE PRESIDENT, 2008

THE RAPID DEVELOPMENT of the tar sands has made climate change a
joke about Everybody, Somebody, Anybody, and Nobody. Everybody
thinks reducing carbon dioxide emissions needs to be done and expects
Somebody will do it. Anybody could have reduced emissions, but
Nobody did. Everybody now blames Somebody, when in fact Nobody
asked Anybody to do anything in the first place.

In the last fifteen years, the federal government has played all four
roles with great élan. In the process it has spent more than $7 billion
(somebody's money) on a half-dozen climate programs with promising
names such as Green Plan, National Action Program, Action Plan, Project Green, the Clean Air Act, and Turning the Corner. Each one has
failed to meet its targets or commitments, let alone to curtail carbon
dioxide emissions that have risen 27 per cent since 1990, the highest
increase of any industrial nation on the planet.

The latest federal strategy, an extravagant carbon burial scheme
known as carbon capture and storage (CSS), will likely meet a similar

fate. In a recent review of what even federal bureaucrats admit are a spectacular succession of "policy catastrophes," the blunt Simon Fraser University economist Mark Jaccard concluded that Canadians like "burning our money to save the planet."

In 2006, Canada's environment commissioner, Johanne Gélinas, laid out the dirty math. She reported that oil and gas production, including tar sands mining, had produced 150 million tons of greenhouse gases in 2004, a whopping 51 per cent increase since 1990. Oil and gas destined for the United States accounted for nearly a third of Canada's increase in total greenhouse gases, approximately the same amount by which Canada failed to meet its Kyoto protocol targets.

Noting that the tar sands had made a major contribution to "increasing greenhouse gas emissions," Gélinas found overall an astounding level of federal neglect and incompetence on climate change and oil production: "Few federal efforts are underway to reduce these emissions and those efforts have had minimal results to date. For its part, the federal government is counting on regulatory and long-term technological solutions... However, it is not leading the way by clearly stating how and to what degree Canada will reduce greenhouse gas emissions when oil and gas production is expected to increase." Gélinas concluded that any further growth in tar sands production would likely cancel out national efforts to lower emissions. Shortly after she produced her damning report, the government fired her.

In 2007, the International Panel on Climate Change outlined the continental effects of rising carbon dioxide emissions from tar sands development and other fossil fuel–burning projects: more heat waves, coastal storms, and freak weather; shrinking glaciers and alpine meadows; less water in our lakes and rivers; fewer wetlands; more forest fires and beetle epidemics; and more human and wildlife diseases as parasites move north to escape the heat. In simple terms, the construction of every new tar sands project contributes to greater economic vulnerability, unreal weather, and chronic water shortages for ordinary citizens.

Part of the carbon problem rests with bitumen's quintessentially dirty character. Unconventional hydrocarbons may sit at the bottom of the energy barrel, but they are another signal of peak oil. Bitumen not only requires more fossil fuels to exploit, its extraction also produces unconventional plumes of pollution and carbon. To make one barrel of oil from the sands, two tons of dirt must be dug up by monster trucks emitting nitrogen oxides (another serious global warmer) and then upgraded at facilities fuelled by natural gas. With all its boiling and steaming, SAGD creates nearly twice as much carbon dioxide as the open-pit mines. As a result, every barrel of bitumen produced from the tar sands creates, on average, three times more carbon dioxide emissions (187 pounds) than a barrel of normal crude (62 pounds). "All unconventional forms of oil are worse for greenhouse gas emissions than petroleum," noted the late Alex Farrell while he was an energy expert at the University of California, Berkeley. "When we face tradeoffs between economics, security and environment, the environment often ends up getting the short end of the stick."

The situation is steadily getting worse as industry mines increasingly poorer grades of bitumen. According to the Petroleum Technology Alliance of Canada, the volume of greenhouse gases per barrel could easily triple from current levels: "Whatever technology is used for inaccessible low quality heavy oil and bitumen deposits, the energy, GHG [greenhouse gases] and water intensities will be higher."

Most statistics on the carbon intensity of bitumen mining don't include the destruction of the boreal forest. Yet the region's hardworking trees and peat bogs now sequester or bank twice as much carbon as a tropical forest. Both open-pit mining and SAGD projects subvert that function by cutting down trees and draining peat bogs. Canada's boreal forest holds 186 billion tons of carbon, and the Mackenzie River Basin protects about 28 per cent of that. Planting giant bitumen mines and factories in the forest is like opening a bank vault to a gang of thieves. Edmonton-based economist Mark Anielski recently calculated that the region's peat bogs, wetlands, and trees stored an estimated $252 billion worth of carbon dioxide. He found that the watershed overall also provided a range of ecological services, such as old-fashioned water-making and filtering, worth $1,064 per acre. The value of these essential services, including carbon saving, greatly outstrips the $99-per-acre market value of bitumen and other minerals in the basin.

Excavating one of Canada's best carbon sinks and weather stabilizers to produce a product with three times the carbon footprint of conventional oil may be an example of global freak economics. Anielski recommended a "more prudent approach" that would safeguard the region's natural capital. No federal or provincial plan has yet emerged.

Many tar sands projects puff out nearly a million tons of carbon dioxide a year. (According to the federal government, a million tons—a megaton—is enough lethal carbon dioxide to fill one million two-storey, three-bedroom homes and suffocate every occupant.) A bitumen upgrader powered by natural gas pumps out 1.3 megatons of carbon dioxide a year. A similar plant that burns bitumen dregs for fuel makes twice as much pollution: 2.6 megatons. A coal-fired plant providing electricity to a tar sands upgrader emits 3.8 megatons of carbon dioxide a year. Syncrude's own large mines and upgraders discharge 14 megatons into the atmosphere annually. Imperial Oil's Kearl project will pollute the atmosphere with 3.6 megatons of carbon dioxide a year, the same amount as 800,000 passenger vehicles. Based on 2002 data, Imperial will represent nearly 2 per cent of Alberta's total greenhouse gas emissions. Regulators have dismissed the amount as "insignificant."

The tar sands are Canada's largest single growing source of carbon dioxide. According to the Tyndall Centre for Climate Change Research, the megaproject accounted for 3 per cent of Canada's emissions in 2004 and will account for more than 16 per cent of the nation's emissions by 2020. That makes the world's largest industrial project a tar nation among nations. If the U.S. Carbon Dioxide Information Analysis Center were to include the project on its list of countries, Alberta's Tar Nation would easily stand out as a significant global polluter. In 2003, the tar sands turned up the global thermostat by adding twenty-five megatons of carbon dioxide to the atmosphere, more than Jamaica or Rwanda. In 2007, Tar Nation dirtied the sky with forty megatons of carbon dioxide. In so doing, it edged out the volume of annual emissions produced by the individual countries of Switzerland, Hong Kong,

Bangladesh, Slovakia, New Zealand, Peru, Ecuador, Cuba, Croatia, Jordan, Sri Lanka, Mongolia, Bolivia, and Panama.

By most estimates, the tar sands will pump out 140 megatons of global heaters by 2020. By that time, the megaproject will be making more greenhouse gases than the current output from the world's volcanoes. The climate-changing gases released by Tar Nation will also exceed those currently produced by the Netherlands, Greece, Argentina, Pakistan, the Czech Republic, Kuwait, Vietnam, Norway, Iraq, North Korea, and Israel. Incredibly, the project will be making as many global warmers as the United Arab Emirates, which holds the fifth-largest proven oil reserve in the Middle East.

The scale of greenhouse gas production in the tar sands raises a key policy question. If Canada exports a dirty fuel that's burned in the United States, which country should be held responsible for polluting the Great Aerial Ocean? The moral answer is probably both. California has already imposed restrictions on dirty oil imports, and the U.S. government's new Energy Independence and Security Act forbids U.S. agencies from spending taxpayers' money on unconventional fuels that create more greenhouse gas emissions. Both Alberta and the government of Canada are arguing for exemptions.

As a way of addressing the problem, industry and government have championed reductions on carbon intensity as opposed to firm caps on carbon production. The emphasis on intensity is a bit of a magic act. While Shell and Imperial marginally decrease the amount of carbon produced per barrel of oil, they wipe out those savings by ramping up oil production. Among economists this problem is known as the Khazzoom-Brookes Postulate.

The postulate dates back to the coal era, when natural resource watchers noted that efficiencies gained by the coal-fired steam engine only momentarily lowered the demand for coal before consumption shot up tenfold. Economists generally agree that increased efficiency in the exploitation of a resource will lead over time to greater consumption, not less. This explains why reductions in energy intensity have yet to translate into reductions in energy demand in Canada, the

United Kingdom, the United States, or anywhere else. The paradox can be found in everyone's driveways, where improved fuel efficiency has added extra cars to the garage and increased the miles driven annually by the average American commuter from 9,500 to 12,000 in the last forty years. (The number of vehicles in Canada has doubled since 1970 to eighteen million and now grows faster than the country's population.) Since 1975, airplanes have worked hard to burn 40 per cent less fuel, but the industry has grown by 150 per cent. As appliances become more efficient, households fill up with electronic gadgets that draw more electricity. The Khazzoom-Brookes Postulate is a rude reminder that energy intensity, like carbon intensity, won't solve a single damn problem without restrictions on energy demand. It also proves, as economist David Brower once noted, that "the promotion of growth is simply a sophisticated way to steal from our children."

THE TAR SANDS are probably the world's largest example of the Khazzoom Brookes Postulate. Although many companies have reduced carbon, water, and energy intensities, exponential growth in oil production has wiped out the small savings. The local community has followed industry's example. The majority of the workforce in the tar sands commutes by plane or drives large vehicles that produce 40 per cent more carbon than an average car. After considering the frightful implications of the postulate, chief economist at CIBC World Markets Jeff Rubin concluded that "reducing total energy consumption must be the final objective to both the challenges of conventional oil depletion and to greenhouse gas emissions." Neither the Alberta government nor the Canadian government wants to accept that reality.

Rapid tar sands development hasn't just multiplied greenhouse gases. It has contaminated the nation's entire environmental track record. A 2005 study of Canada's performance on issues such as sulfur dioxide emissions and the generation of nuclear waste revealed that it ranked twenty-eighth out of twenty-nine Organisation for Economic Co-operation and Development (OECD) nations. Moreover, Canada's performance on many environmental indicators was getting worse. The study cited:

"Increasing water consumption, increasing energy consumption, increases in nuclear and hazardous waste, higher greenhouse gas emissions; higher numbers of endangered species; declining fish populations; higher commercial fertilizer use; more livestock; more timber logged; more motor vehicles; more kilometres traveled by road, higher population." Air-quality monitoring was so haphazard that a North American progress report on acid rain left the maps of Canada blank because of insufficient data between 2000 and 2004. A 2007 federal report, *Canadian Environmental Sustainability Indicators*, confirmed that both smog levels and water quality are getting worse: "Pressure on Canada's environment is steady or increasing."

Most Canadians understand that the carbon volcanoes in Fort McMurray, Peace River, and Cold Lake, along with rapid population growth, forced the country to abandon its international obligations under the Kyoto Protocol. The country's failure to honour the accord has been a source of international embarrassment and much green angst, but no Canadian should weep long or hard on that account. Kyoto was a poor agreement, drawn up by government bureaucrats and non-governmental organizations with no real understanding of peak oil or the Khazzoom-Brookes Postulate. The protocol was also the product of what one critic calls "the monomaniacal fixation on a consensual global solution" to climate change.

The highly complex agreement, written in oblique language, committed industrial nations to minimal targets: 5 per cent reductions by 2012. (Canada agreed to 6 per cent, then made no effort other than burning money to meet it.) Every climate change expert knows Kyoto's itsy-bitsy targets can't and won't prevent runaway global warming.

Second, the protocol included a number of voluntary "flexible instruments" that allow carbon-rich nations to sell their dirty laundry to carbon-poor nations. In the process, the high-carbon crowd can earn green credits or bonus points by building coal projects in low-carbon nations. Hermann Scheer, a cogent German energy analyst, notes that this Byzantine scheme has created in Europe "a bureaucratized and correspondingly inflexible system of investment controls." Instead of reducing carbon emissions and moving the economy to renewable energy, Kyoto did a fine job of driving up costs by increasing "the number of lodgers and boarders in the energy system." Last but not least, Kyoto tried to legitimize "an unsustainable condition," argues Scheer, by permitting carbon pollution that could have been avoided through direct investments in renewable energy or fossil fuel conservation.

Unfortunately, in carbon capture and storage, the Canadian government has come up with a plan that may be worse than Kyoto, though David Keith, who holds the Canada Research Chair in Energy and Environment at the University of Calgary, thinks CCS is the only national option left: "Given the dominance of the fossil fuel industry and our engineering experience, CCS is necessary if you want to preserve the Alberta economy."

With great fanfare and lots of adjectives—*tough* was the favourite—the Canadian government announced in March 2008 that it would create an innovative class of carbon undertakers subsidized by taxpayers. The proposal is strictly funereal: all new coal-fired plants and tar sands projects will capture their carbon dioxide, tidily compress the elusive climate changers (kind of like stuffing a body into a suitcase), and then inject the waste deep under the prairie by 2018. Following that, alert federal or provincial civil servants will monitor the carbon for thousands of years. With the exception of one carbon-recycling scheme used for oil recovery in the Williston Basin near Weyburn, Saskatchewan, no infrastructure currently exists to bury carbon. To inject twenty megatons (an amount equivalent to the annual tailpipe exhaust of four million vehicles) will cost anywhere from $10 billion to $16 billion. In 2008, shortly after the Task Force on Carbon Capture and Storage requested $2 billion in federal money to explore how to effectively bury just five megatons, the Alberta government hospitably created a $2-billion public fund to subsidize a host of private CCS projects. One of the first beneficiaries, Capital Reserve Canada Ltd., is a company partly owned by former Alberta Premier Don Getty.

Alberta and Ottawa, the key beneficiaries of rapid tar sands expansion, favour CCS as the best way to create a "sustainable energy superpower." The technocrats argue that Canada's largest source of greenhouse gas pollution demands "Canada's largest single sector CO_2 mitigation option." In other words, a big problem demands one big solution. As the Alberta government notes in its *Oil Sand Facts*, "bigger is better." The province even has former Syncrude CEO Jim Carter directing a provincial council on the future of carbon capture. The task force on CCS argues that the technology will allow Canada "to build on its existing energy infrastructure and its fossil energy endowment while managing the associated GHG emissions." It estimates that CCS has the potential to reduce Canada's carbon storm by 60 per cent to 70 per cent by 2050. Proponents also point out that Canada has a major carbon graveyard in the Western Canadian Sedimentary Basin: empty oil and gas fields, depleted reservoirs, and deep salty aquifers. Conveniently, CCS will extend the life of fossil fuel business.

But the economics of CCS are deadly. According to the International Panel on Climate Change, the cost of capturing just one ton of carbon ranges anywhere from $25 (U.S.) to $115 (U.S.). The Canadian Library of Parliament reported in a 2006 research paper that the act of capturing the carbon can eat up nearly 30 per cent of the energy produced by a power plant or a tar sands project. "These parasitic power losses mean you have to mine and burn more coal [or natural gas or bitumen] in order to cover the cost of burying the emissions," explains Dave Hughes, a retired coal specialist with Natural Resources Canada. Sticking a carbon burial unit onto a coal-fired plant typically raises the costs of electricity production by anywhere from 37 per cent to 91 per cent. Hughes reasons that coal and tar sands companies like the idea because it will keep them in business longer, "but it's not a great long-term strategy for the human race." A 2007 German study published in the *International Journal of Greenhouse Gas Control* (yes, there is such a journal) found that renewable forms of energy such as wind and solar could be developed more quickly and in the long term be cheaper than CCS. Canada, of course, hasn't done such a study yet. Scientists also point out that the CCS is largely untested. "In full scale the technology only exists in the imaginations of the people developing it," says

Swedish scientist Anders Hansson. "It's overly optimistic to place such great faith in it."

Critics say that if the Canadian and Alberta governments were really serious about reducing carbon emissions, they would pick the lowest fruit first: fugitive emissions from the upstream oil and gas sector. According to the Petroleum Technology Alliance Canada (PTAC), pipelines, well heads, and gas plants leak nearly ninety-nine megatons of carbon dioxide and methane into the atmosphere every year—14 per cent of Canada's total emissions. Moreover, studies have shown that industry computer models grossly underestimate how bad the leaks are. A 2005 study by the Alberta Research Council, for example, found that a sweet-gas plant fumigated the neighbourhood with 1,224 tons of methane a year, ten times more than the plant had estimated. (Methane is a much more destructive greenhouse gas than carbon dioxide.)

Simple regulation and air monitoring could force industry to plug all the holes in its equipment. Unlike Alberta, European countries routinely audit and compel repairs to their refineries. The main obstacle to reducing air pollution and conserving nearly a billion dollars' worth of hydrocarbons a year, says PTAC, is Canada's tar sands factor, a frantic demand for more oil and gas production that "competes for capital and people."

University of Manitoba professor and energy expert Vaclav Smil calls CCS "a third rate option," or a "General Motors approach to living." Faced with demands to lower pollutants from its cars in the 1970s, GM decided not to reduce nasty stuff such as nitrogen oxides. It patched onto its vehicles an expensive three-way catalytic converter that produced lots of heavy metals instead. Honda, in contrast, decided to make a combustion engine that eliminated the pollution altogether. Smil argues that Honda's decision to minimize bad outputs "should be the guiding principle of any intelligent, far-sighted rational design." Today, Honda is a thriving company. GM is not.

Smil calculates in his recent book *Energy Myths and Realities* that it would take a geo-engineering endeavour of unprecedented magnitude to compress, transport, and store just 15 per cent of the world's carbon dioxide emissions. He writes, "We would have to put in place a gathering, compression, transportation and storage industry whose annual volume through put would be slightly more than twice that of the annual volume through put of the world's crude oil industry with its immense networks of wells, pipelines, compressor stations, tankers and above and underground storage...Needless to say, such a technical feat could not be accomplished within a single generation." Smil's arithmetic suggests that lofty plans to bury 50 per cent of Canada's carbon dioxide emissions by 2050 is a pipe dream. ccs, he argues, is part of the same thinking that gave us the energy spectacle of "a 50-kg female driving a 3,000-kg suv in order to pick up 1-kg carton of milk."

The safety issues related to ccs have also escaped serious examination. The Canadian government wants to hide tar sands pollution deep under the prairies, in salty aquifers near cities such as Regina and Edmonton. But the drilling of 350,000 oil and gas well sites has made Western Canada one of the most perforated landscapes on Earth. Methane and other hydrocarbons already migrate from poorly sealed and concreted wells into groundwater and people's homes. (In some gas fields, 80 per cent of the wells leak.) Even ccs proponents admit that carbon dioxide injected deep underground could find its way back to the surface after an earthquake or via groundwater channels.

Bruce Peachey of New Paradigm Engineering doesn't think anyone has honestly assessed the risk of exposing the public to massive clouds of carbon dioxide, a gas heavier than air. In an open letter he wrote in 2008, he noted, "If a blowout, or significant leak, were to occur, high ground level CO_2 concentrations would prevent anyone from approaching the well without an air pack and even equipment engines would need air supplies." The International Panel on Climate Change recently calculated that a even if a minor continuous leak of carbon dioxide occurred, it would offset the benefits of storing carbon. A leakage rate of 0.1 per cent, for example, would empty a carbon graveyard in less than six thousand years.

Creating an energy-intensive burial system to hide a problem that could be solved by conserving fossil fuels is morally bankrupt. ccs is a last-ditch survival effort that defies economics and shirks logic. It extends the pretence that carbon is not connected to dirty oil and that business as usual in the tar sands is sustainable. It assumes that naive taxpayers will pick up the multibillion-dollar tab and that neighbouring communities will gladly assume the risks of living downwind from potentially leaky CCS cemeteries.

Fossil fuel exports now account for 10 per cent of national emissions, and dirty oil from the tar sands has helped to create a 109 per cent increase in climate-warming gases from SUVs, vans, and trucks since 1990. Canada now has one of the highest rates of energy consumption and carbon dioxide production per person in the world. In 2006, the U.S. Energy Information Administration rated Canada as the third most energy-intensive and the fourth most carbon-intensive economy among OECD countries. Canada's proposed solution to this carbon fiesta is a third-rate burial service funded by taxpayers that won't make a dent in carbon emissions until 2020.

Nearly thirty years ago, U.S. physicist Albert Bartlett wrote a paper called "Forgotten Fundamentals of the Energy Crisis." Though it remains largely unread among politicians, it offers a perceptive analysis of the tar sands frenzy: "We must realize that growth is but an adolescent phase of life which stops when physical maturity is reached. If growth continues in the period of maturity it is called obesity or cancer. Prescribing growth as the cure for the energy crisis has all the logic of prescribing increasing quantities of food as a remedy for obesity."

At the moment, Canada's official prescription calls for more growth, more dirty oil, and more pollution. Like an increasing number of people around the globe, George Monbiot, the combative *Guardian* columnist and author of *Heat,* pointedly accuses Canada, as "one of the planet's most polluting nations," of choosing badly: "You could scarcely do more to destroy the biosphere if you tried."

He is right.

TEN

NUKES FOR OIL!

....

"Nuclear power based on fission is potentially larger than
the fossil fuels, but it also represents the most hazardous industrial
operation in terms of potential catastrophic effects that has
ever been undertaken in human history."

MARION KING HUBBERT, SHELL GEOPHYSICIST, 1956

IN 1956, MANLEY NATLAND had the kind of energy fantasy that the tar
sands invite with predictable regularity. As the Richfield Oil Company
of California geologist sat in a Saudi Arabian desert watching the sun
go down, it occurred to him that a nine-kiloton nuclear bomb could
release the equivalent of a small, fiery sun in the stubborn Alberta tar
sands deposits. Detonating the bomb underground would make a massive hole into which boiled bitumen would flow like warmed corn syrup.
"The tremendous heat and shock energy released by an underground
nuclear explosion would be distributed so as to raise the temperature of
a large quantity of oil and reduce its viscosity sufficiently to permit its
recovery by conventional oil field methods," Natland later wrote. He
thought that the collapsing earth might seal up the radiation, and the
bitumen could provide the United States with a secure supply of oil for
years to come.

Two years after his desert vision, Natland and other Richfield Oil representatives, the Alberta government, and the United States Atomic Energy Commission held excited talks about Project Cauldron, which planners later renamed Project Oil Sands. Natland selected a bomb site sixty-four miles south of Fort McMurray, and the U.S. government generously agreed to supply a bomb. Richfield acquired the lease site. Alberta politicians celebrated the idea of rapid and easy tar sands development, and the Canadian government set up a technical committee. *Popular Mechanics* magazine enthused about "using nukes to extract oil."

Edward Teller, the nuclear physicist and hawkish father of the hydrogen bomb, championed Natland's vision. In an era when nuclear proponents got giddy about nuclear-powered cars, Teller regarded Project Cauldron as another opportunity to hammer the threat of nuclear swords into peaceful ploughs. "Using the nuclear car to move the fossil horse" was a promising idea, the bomb maker wrote.

Chance, however, intervened. Canadian Prime Minister John D. Diefenbaker didn't relish the idea of nuclear proliferation, or of the United States meddling in the Athabasca tar sands. The Soviets had experimented with nuking oil deposits only to learn that there was no market for radioactive oil. The promise of cheaper conventional sources in Alaska also lured Richfield Oil away from Project Cauldron.

The moment passed for Natland. But the idea of using a nuclear car to fuel a hydrocarbon horse never really died, and these days some new scheme to run the tar sands on nuclear power emerges weekly with great fanfare. The CEO of Husky Energy, John Lau, seems interested, and Gary Lunn, the federal minister of natural resources, says he's "very keen," adding that it's a matter of "when and not if." Roland Priddle, former director of the National Energy Board and the Energy Council of Canada's 2006 Energy Person of the Year, speaks enthusiastically about the synthesis "of nuclear and oil sands energy," as does Prime Minister Stephen Harper. Bruce Power, an Ontario-based company, has proposed four reactors at a cost of $12 billion for tar sands production in Peace River country. France's nuclear giant Avera wants to build a couple of nukes in the tar sands too. Saskatchewan, an Alberta wannabe, has proposed two nuclear facilities: one near the tar sands and one on Lake Diefenbaker. Employees of Atomic Energy of Canada Ltd. (AECL), a federal Crown corporation that designs and markets CANDU reactors, told a Japanese audience in 2007 that "nuclear plants provide a sustainable solution for oil sands industry energy requirements, and do not produce GHG emissions."

If realized, these latest atomic visions for the tar sands would make Canada the only developed country in the world to employ nuclear power to accelerate the exploitation of carbon-rich fossil fuels. The notion has stumped even the fine minds at the esteemed journal *Petroleum Economist*: "Building a plant that makes clean energy in order to produce more dirty oil belies Ottawa's claim that a nuclear plant in the oil sands patch would be built for environmental reasons." It actually belies much more.

Geologists have always been ambivalent about accepting nuclear energy as a substitute for fossil fuels. In his now-famous 1956 lecture on the two power sources delivered in Houston, Texas, Shell geophysicist Marion King Hubbert predicted that U.S. fossil fuel production would peak in the 1970s and make the United States increasingly dependent on foreign supplies and politics. With humankind's use of oil a "non-repetitive blip" in history, King initially thought that nuclear power might be a passable substitute. But nuclear accidents, nuclear waste, and the proliferation of weapons of mass destruction convinced him that nuclear energy wasn't the answer. He later threw his weight behind solar energy, "the biggest source of energy on this earth," advocating that oil and gas companies throw their weight behind it too.

But that's not what has been happening. In sunny Alberta, nukes for oil are being celebrated these days as some sort of magic bullet for carbon pollution as well as for rapid depletion of natural gas supplies. Natural gas now fuels rapid bitumen production, and it takes approximately 1,400 cubic feet of natural gas to produce and upgrade a barrel. (Incredibly, that's equal to nearly a third of the barrel's energy content.) The tar sands are easily Canada's biggest natural gas customer. They burn the blue flame to generate electricity to run equipment and facilities, they convert it as a source of hydrogen for upgrading, and they use it to heat water. SAGD operations, which need anywhere from two to four barrels of steam to melt deep bitumen deposits, are super-sized natural gas consumers. Thanks to the unexpectedly low quality of many bitumen deposits, SAGD requires more steam and therefore more natural gas every year.

In 2006, the *Oil & Gas Journal* noted sadly that Canada had only enough remaining natural gas to recover 29 per cent of the bitumen in the tar sands. The North American Energy Working Group (NAEWG) reported similar findings that year at a meeting in Houston, Texas. If the tar sands produced five million barrels a day, the group said, oil companies would consume 60 per cent of the natural gas available in Western Canada by 2030. Even the NAEWG found that level of consumption "unsustainable and uneconomical." As one Albertan recently observed: "Using natural gas to develop oil sands is like using caviar as fertilizer to grow turnips."

Thanks to Alberta's cavalier approval process, the oil patch is now building $4 billion worth of SAGD projects every year. That means by 2015 the tar sands will consume 16 per cent of the nation's gas supply, enough of the blue flame to warm approximately twelve million homes twenty-four hours a day. And many observers think those estimates are far too conservative. Armand Laferrère, the president and CEO of Avera Canada, estimates that the tar sands industry could commandeer 92 per cent of Canada's natural gas supply by 2030. That doesn't leave much gas for winter home-heating.

Wayne Henuset, a Calgary oil man who became a climate change believer after a hurricane nearly levelled his Florida dream home, issues similar warnings. By 2015 ("and that's not very far away"), Henuset estimates the tar sands will soon burn more than three billion cubic feet of gas a day, "more natural gas than all the rest of Alberta uses now," and more natural gas than would be delivered by the proposed $16-billion Mackenzie Valley gas pipeline. Henuset says it's stupid "to squander precious and declining reserves of natural gas to make oil in the oil sands. That's simply like burning gold to make coal." His answer is nuclear power.

Bill Gwozd, a vice-president of gas services for Ziff Energy Group, told the *Calgary Herald* in 2008 that Canadians will soon be asked "whether it's more important to run our Nintendos, cellphones and laptops than [to] have enough gas to produce oil for valuable export markets. It's just shock and awe that there's no government policy [on natural gas depletion]."

The promise of taking the carbon bite out of bitumen also appears on the nuclear agenda. Wayne Henuset says that just one ARC1000 CANDU reactor could offset nearly 500 million tons of carbon dioxide a year and "result in major oil sands projects having a near zero increase in greenhouse gas emissions over its life expectancy." In 2007, a group of researchers at the Massachusetts Institute of Technology evaluated the effectiveness of three kinds of nuclear reactors for tar sands production. They concluded that nuclear energy was not only two to three times cheaper than natural gas but would reduce greenhouse gas pollution from a plant producing 100,000 barrels of bitumen a day by about 100 megatons annually. The only drawback, reported *Nickle's Daily Oil Bulletin*, is that the nuclear industry requires "some kind of market pull and public acceptance." Stephen Harper has begun work on the public acceptance angle by calling nuclear power a "no-emissions" wonder.

But the analysts and salespeople championing nuclear power as a carbon fighter and natural-gas conserver generally forget to add a few critical caveats. For starters, nuclear power remains the most expensive and capital-intensive fuel on the planet. The business community has never been fond of such highly centralized technology because it tends to behave like a Soviet commissar, with little financial accountability. (Margaret Thatcher tried to privatize England's nuclear power plants, but no entrepreneurs were dumb enough to volunteer for the money-losing opportunity.) Plagued by cost overruns and technological failures, public utilities served by nuclear power carry some of the world's highest debt loads. In fact, no nuclear power plant has been built in Canada on budget or without taxpayers' money. Électricité de France, which receives 85 per cent of its power from nuclear reactors, is among the most debt-ridden companies in the world. So is Ontario Hydro.

Contrary to industry claims, nuclear power isn't adept at reducing greenhouse gas (GHG) emissions, either. It takes a lot of fossil fuels to mine and enrich uranium ore and then turn it into a fuel rod. As a consequence, nuclear power indirectly emits approximately 250,000 tons of carbon dioxide a year. In 2006, the German Institute for Applied Ecology compared the cost and GHG emissions of nuclear power with a variety of other energy sources and found that nuclear electricity was not the winner. Institute scientists noted that the "net CO_2 emissions of electricity from gas-fired ICE cogeneration plans are *lower* than the CO_2 emissions of electricity produced in nuclear power plants...All in all renewable electricity and electricity efficiency have lower GHG emissions than nuclear electricity." A nuclear power plant might have to run for a decade before it could legitimately call itself "free" of carbon dioxide. In any case, the Institute for Energy and Environmental Research reported in 2006 that it would take the construction of a new nuclear power plant every two weeks until 2050 to truly make a dent in global greenhouse gas emissions.

Next comes the water issue. Nuclear plants overheat without regular baths of cool water. (This explains why current proposals have placed nuclear reactors on the Peace River, one of Alberta's longest rivers, or Lake Diefenbaker, the source of 40 per cent of the water for Saskatchewan.) The Darlington and Pickering facilities in Ontario require approximately two trillion gallons of water for cooling a year, about nineteen times more water than the tar sands use. In fact, water has become an Achilles heel for the nuclear industry. Recent heat waves in Europe and the United States either dried up water supplies or forced nuclear plants to discharge heated wastewater into shallow rivers, killing all the fish. Global warming, in its inimitable way, has highlighted the limits of nuclear power.

Waste poses another conundrum. Although the tailings ponds are predicted to leave a mess for nearly a thousand years, nuclear power will leave behind a glowing garbage pile for a hundred thousand years. Canada still has no answer to that problem.

Finally, the nuclear power industry in Canada has a history of secrecy and regulatory neglect. Leaks at the Chalk River reactor in Ontario, violations in mine safety, and a legacy of community public health problems and lung cancers among uranium workers have never been openly investigated. The 2007 firing of the head of the Canadian Nuclear Safety Commission for trying to keep things safe indicates a trend as reassuring as the performance of Alberta's ERCB.

With all of these problems, it's easy to see why the industry embraces dirty oil as its saviour. A rather grandiose 2005 proposal to plant a nuclear complex (the world's largest) on Cree Lake, just 146 miles east of Fort McMurray in the middle of Canada's uranium mining belt, spelled out the industry's interest in bitumen: "The simultaneous exploitation of the uranium and oil sands resources, with nuclear process power producing millions of barrels per day of synthetic crude oil, is an economic roadmap to sustained growth in both industries."

While most countries want to employ nuclear power to retire or replace dwindling fossil fuels, Alberta sees it as a way to produce more bitumen, including a hydrocarbon even dirtier than tar in what geologists call the Carbonate Triangle. The triangle is a 27,000-square-mile area lying below the tar sands that contains about 26 per cent of the province's bitumen. Instead of being mixed with sand, this bitumen is locked in dense limestone and heavily karsted rock. Extracting it is an energy-intensive and water-gulping proposition much like that in Colorado's oil shales. To date, Royal Dutch Shell has invested $400 million in the region.

According to an AECL presentation delivered in Oarai, Japan, in 2007, the best way to extract bitumen from limestone is with electrical heaters in thousand-foot-long vertical tubes. After three years of intensive cooking by these fancy electrodes at up to 1100 degrees Farenheit, the rock will surrender its gas and light oil. Given the hefty electricity bill, only nuclear power would make carbonate bitumen remotely economical at a production cost of $30 a barrel. By some accounts, extracting Alberta's carbonate deposit could require as many as fourteen nuclear reactors, but Shell remains mute on its future plans.

The other quiet yet key driver for nuclear power is the U.S. water and energy crisis. Unlike Environment Canada, a neutered organization with little credibility, the U.S. Department of Energy understands that you can't make energy without water. Water, like oil, is a diminishing resource south of the border. In a 2006 study called "Energy Demands on Water Resources," the department painted a bleak picture of the failing marriage of water and energy in the U.S. economy. It reported that thermoelectric power generation (80 per cent of U.S. electricity comes from fossil or nuclear power generation plants) accounted for 40 per cent of all freshwater withdrawals, mostly for cooling, and that the faucet was running dry: "Low water levels from drought and competing issues have limited the ability of power plants to generate power."

To further highlight the critical interdependence of water and energy, the report showed that the states most vulnerable to water shortages, such as Texas, California, and Louisiana, have been centres of intense oil and gas exploitation. The report warned that the ability to easily expand freshwater availability for energy production in the United States "may be limited" and that water shortages could well contain the growth of the nuclear industry. Canada, the most selfless of neighbours, now offers an ideal solution. Why not locate nuclear power stations on Arctic-bound rivers in order to send both electricity and bitumen south?

This enticing scenario probably explains why TransCanada Corporation, a nuclear power proponent as well as a Bruce Power partner, wants Alberta's electricity grid to be connected to the western United States. Such a connection would allow the province's new nuclear power plants, when not providing electricity, steam, or hydrogen to the tar sands, to supply "clean energy" to Los Angeles. (The suggestion that Albertans might oppose such schemes, says TransCanada's CEO Hal Kvisle, is the kind of thinking that is "very short-sighted and ill founded.")

Such powerful visions also explain why the U.S. Federal Energy Regulatory Commission is now rewiring continental electricity standards. (Although many U.S. states have strongly objected to this centralized push, Canadian regulators have signed on like sheep.) The Canadian Electricity Association argues that the best way to build energy security is not through silly schemes for energy independence or conservation but through the integration of U.S. and Canadian energy markets along with "opportunities to promote nuclear facilities."

In recent years, given the threat of climate change and peak oil, highly respected environmental activists such as Stewart Brand have advocated for nuclear power. They view it as one slice of the energy-solution pie. James Lovelock, the author of the Gaia Hypothesis, has routinely berated anti-nuke greens for their knee-jerk attacks on nuclear energy: "Even if they were right about its dangers, and they are not, its worldwide use as our main source of energy would pose an insignificant threat compared with the dangers of intolerable and lethal heat waves and sea levels rising to drown every coastal city in the world... Civilization is in imminent danger and has to use nuclear—the one safe, available, energy source—now or suffer the pain soon to be inflicted by our outraged planet."

But building nukes to increase fossil fuel production would probably strike Lovelock as a suicidal endeavour. Nevertheless, the tar sands have opened that Pandora's box. The Alberta Research Council, a corporation owned by the government of Alberta, has joined forces with the Idaho National Laboratory, U.S. Energy's main nuclear think tank, to work out the essential details of what the lab's associate director, Bill Rogers, calls "our energy security goals." Rogers has also described the joint venture of bitumen production and nuclear experimention as a "marriage made in heaven." (In 2007, his busy lab reported that "in the near-term hydrogen from nuclear energy will be used to upgrade crude.") The Cambridge, Massachusetts, office of Shaw Stone & Webster Management Consultants is calling Alberta universities to inform faculty members about "the potential environmental benefits of using nuclear energy in the oil sands" in preparation for "a public outreach initiative." Project Cauldron has resurfaced.

Kjell Aleklett and two graduate students at the Swedish Uppsala Hydrocarbon Depletion Study Group estimated in 2006 that it would take at least seven years before CANDU reactors could be built to power SAGD operations producing 150,000 barrels of bitumen a day. Given the inadequacy of North America's gas supply, Aleklett believes it will be impossible for SAGD production to reach the projected five million barrels a day by 2030 without lots of nuclear plants providing the electricity and steam. He calculates that nukes might be needed to power energy-intensive "large scale CO_2 sequestration techniques" as well. The study concludes that the depletion of natural gas, combined with the onerous task of building nuclear power plants, illustrates "the great difficulties of rapidly expanding the oil sands industry of Canada in any practical way." The Canadian Parliament estimates it would take twenty nuclear reactors to replace natural gas as a fuel source in the tar sands by 2015.

The push to build nuclear plants for accelerated tar sands production can be felt across the continent. Due to peak oil and climate change, uranium oxide prices have blossomed from $10 a pound to more than $140 a pound in the last decade. Civilian reactors now mine diluted enriched materials from military programs to supply 40 per cent of their needs. Like the end of cheap oil, the end of cheap uranium has created a careless boom. In Northern Ontario, Aboriginal leaders have gone to jail rather than let uranium prospectors lay claim to the future of their land. In New Brunswick, the rights of landowners and the rights of uranium prospectors clash routinely. Throughout the west, from Alberta to Utah, companies are staking claims to iconic landscapes with grades of uranium ores as poor and bottom of the barrel as bitumen.

While nuclear advocates such as Bruce Power push for facilities in both Alberta and Saskatchewan, nuclear entrepreneurs claim the future may well belong to portable nuclear generators the size of a bathtub. Their fission reactor requires no moving parts or human operator. In any case, the pursuit of bitumen in sand or rock seems indelibly linked to the proliferation of nuclear energy in Canada, a costly chain reaction that will radiate more problems than solutions.

ELEVEN

THE MONEY

....

"This province will join the top five to ten oil producers
in the world and we have record-keeping that worked for
Third World countries in the 1960s."

EVAN CHRAPKO, ENTREPRENEUR, *EDMONTON JOURNAL*, 2007

AFTER OIL PRICES shot to $75 a barrel in 2006, ordinary Albertans
started to ask angry questions about where the money was going. As the
collective owners of the tar sands, many wondered if the province, bur-
dened by a $7-billion infrastructure debt, was getting its fair share of the
pie. In oil-rich states, governments typically charge royalties on the
resource both to save for a rainy day and to keep the companies competitive. A royalty is also a reminder that oil is a one-time public treasure.
Yet Alberta hadn't bothered to publicly review its outdated royalty rates
since 1992. Energy Minister Greg Melchin promised Albertans a study
but then declared, without it, "We've decided we get our fair share in
Alberta on both conventional oil and oil sands." He later explained that
an internal technical review supported the decision and left it at that. In
other words, an emerging energy superpower doesn't have to provide
the public with graphs or charts supporting its decisions.

The inquiries persisted, particularly after some members of the province's legislature admitted that they had never seen a review and doubted that the government had done one. Finally, Alberta Premier Ralph Klein tried to silence the growing impertinence with a remarkable sound bite: "We do get our pound of flesh. I don't know if it [the review] was completed or not, nor do I give a tinker's damn. I've always been satisfied that our royalty regime is proper and fair."

But the premier lied and lied boldly, as did Melchin and his successor as energy minister, seasoned oil man Mel Knight. The public record damningly shows that Alberta's royalty regime is neither proper nor fair. Although government studies and reviews had highlighted yearly billion-dollar losses, Klein's government did not share those reports with citizens. Since that time a public panel and the province's auditor general have both revealed that Alberta's system for collecting oil and gas royalties is of such inferior quality it would disgust most corporate accountants. A comparative study of royalty rates charged by the U.S. government also documented what Klein wouldn't: that Alberta's rates were (and remain) among the world's lowest. Nonetheless, the Alberta government has maintained Klein's "tinker's damn" approach. As a result, the province makes much less from its dirty oil than do Norway, Alaska, New Mexico, or even Louisiana. It also makes much less than the Canadian government does. According to a 2005 report by the Canadian Energy Research Institute, Ottawa will rake in $51 billion in corporate taxes from the tar sands between 2000 and 2020, while Alberta will take home only $44 billion.

The truth about Alberta's missing billions emerged piecemeal, like quarters and dimes discovered under an old couch cushion. One of the first revelations appeared in a series of articles written by Gordon Jaremko while he was a business reporter at the *Edmonton Journal*. Jaremko noted that Alberta's share of the energy pie reached a peak of 40 per cent of total oil and gas revenues in the province in 1978, during Premier Peter Lougheed's era. Under Klein, the province's total share of oil and gas revenues fell like lead to 22 per cent. Between 2001 and 2006, as oil and natural gas prices rose, Alberta's share dropped to lows of 15 per cent. In fact, Klein's government repeatedly failed to secure its own target of 25 per cent. In a province where everything is about the money, Klein oddly pretended that the money didn't matter.

The dramatic decline owed much to Alberta's unusual 1 per cent tar sands royalty, which followed the 1996 *Declaration of Opportunity*. The fantastic deal allowed companies to pay 1 per cent against gross revenue until they had recovered all of a project's multibillion-dollar setup costs plus a return on investment. At that point, a 25 per cent royalty kicked in. But the 1 per cent royalty also served as an incentive to expand mines or SAGD projects, thereby postponing higher royalty payments to Albertans. At $20 a barrel, the scheme made some economic sense, but at $75 a barrel it gave companies a licence to print money at the province's expense. Not surprisingly, the 1 per cent royalty rate (a rate Venezuela abandoned long ago) collapsed tar sands revenue by 32 per cent between 1996 and 2006. Oil sands production grew by 123 per cent during the same time. Alberta actually made more money from video lottery terminals than it did from the tar sands between 2001 and 2004.

Lougheed got the debate going when he publicly proclaimed in 2006 that Albertans weren't getting their fair share and also that rapid tar sands development was a disgrace. "I was just up there on a trip, just helicoptering around, and it is just a moonscape," he told *Policy Options* magazine. "It is wrong in my judgment, a major wrong, and I keep trying to see who the beneficiaries are. Not the people in Red Deer, because everything they have got is costing more. It is not the people of the province, because they are not getting the royalty return that they should be getting, with $75 oil."

Although Alberta politicians and officials lied about the state of royalties at home, they declared the truth south of the border. Murray Smith, Alberta's ambassador to the United States, was open in telling the Austin, Texas, crowd in 2006 that "the royalty structure for oil sands is we 'give it away' at a one per cent royalty structure." Roland Priddle, former director of the National Energy Board, endorsed the "give it away" model of business at another Austin energy gathering, this one sponsored by the Canadian consulate general. Referring to Chevron

Corporation's purchase of a tar sands lease with a proven 7.5 billion barrels for $70 million, Priddle asked, "Where else can you purchase in place oil (well, bitumen) for one cent a barrel?" The answer is nowhere.

Late in 2006, economist Robert Mansell at the University of Calgary quietly added to the hue and cry with a critical paper on "Energy and the Alberta Economy." He noted that provincial royalties had dropped precipitously to 15 per cent of total oil and gas revenues and that with the depletion of conventional supplies, combined with the 1 per cent tar sands program, Albertans would earn less and less from their hydrocarbons in the future. He also confirmed that many tar sands companies paid royalties not on upgraded synthetic crude but on raw bitumen, a product worth half as much. As a consequence, Albertans would soon be making $5 billion a year in royalties from their hydrocarbon bonanza instead of the previous average of $8 billion, and all during a boom in oil prices.

After Ralph Klein's reluctant retirement in 2007, Alberta's new premier, Ed Stelmach, set up a public panel to review royalties, in response to growing accusations that Albertans were being cheated. ("When this royalty review is completed, it will become very clear that Albertans have been well served by the system, and they will continue to be well served by the system," vowed Stelmach's new energy minister, Mel Knight.) The government-appointed panel included academics and well-known members of Alberta's business community: Bill Hunter, a former CEO of a forestry company; Evan Chrapko, a millionaire technology developer and chartered accountant; Sam Spanglet, a former Shell executive; Kenneth McKenzie, a professor of economics at the University of Calgary; André Plourde, a professor of economics at the University of Alberta; and Judith Dwarkin, chief economist for the Ross Smith Energy Group.

The panel conducted a series of public hearings, where they got an earful from tar sands developers. The 1 per cent royalty system worked well, said the companies, and the industry was creating jobs. Neil Carmata, vice-president of Petro-Canada, spoke for many executives: "It is clear oil sands is a key element of Petro-Canada's future—as it is for Alberta, and for Canada as an energy super power in the making. This is not the time to change royalties that discourage investment in the very projects that could fuel Alberta's economic future. We should not be hampered in our shared goal of generating added value to the oil sands or our search and development of oil and gas in a conventional environment that is moving to non-conventional." Ordinary citizens such as engineer Rick McCosh offered a different take: "Oil sands projects no longer need incentives to be viable. In fact, the lucrative nature of the current royalty regime is actually contributing to the unrestrained oil sands development now being experienced in Alberta." Straightforward questions about tar sands revenue put to Alberta's Department of Energy couldn't be answered because the department didn't have the data.

As Albertans were speaking up, the U.S. Government Accountability Office (GAO), a watchdog of Congress, shed more light on the true value of Klein's "tinker's damn." After examining royalty fairness in the light of higher oil prices, the GAO concluded that the U.S. government, like Alberta's, secured a ridiculously small share of oil resources. In fact, the GAO report found that Washington, D.C., made less from oil than did Angola, Australia, the U.K., Egypt, and Trinidad and Tobago. The only jurisdiction that fared worse was Alberta. The province earned less for its citizens than did several U.S. states, including Wyoming, Texas, Oklahoma, and California.

In September 2007, Alberta's royalty review panel released its 104-page report, *Our Fair Share*. The paper, a model of economy and clarity,

**GOVERNMENT SHARE OF INDUSTRY OIL REVENUES
COLLECTED THROUGH ROYALTIES**

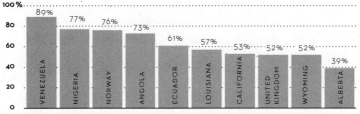

Source: United States Government Accountability Office. Data supplied to the Alaska State Legislature, 2006 (GAO-07-676R).

confirmed the GAO findings and much more. For starters, the panel found "an absence of accountability" so profound that "Albertans cannot determine whether their interests are being well served or whether or not the royalty system is performing as intended." The report showed that the province had failed to accurately measure production data over time, failed to review royalty regimes in an open manner, and failed to collect royalties efficiently. Alberta had "the lowest government takes" of almost any government in the world. Moreover, the panel calculated, the province was failing to collect $2 billion a year in revenue. In 1996, the Alberta government had agreed that the government take from the tar sands should be around 60 per cent with oil at $20 a barrel. The actual government take has averaged around 47 per cent in an era when a barrel of oil has bypassed $100. "Failure to collect the fair share of Albertan's royalties because of overly accommodative rules and under funded administration is a false economy," stated the report. "Imagine the repercussions if the income tax system experienced such a drift and nobody knew or nobody seemed to give a tinker's damn." The tar sands royalty and tax regime, it concluded, "no longer reflects a fair share and balance between owners and the growing number of producers."

The panel's solutions were businesslike. Its report recommended that tar sands royalty rates be increased to 33 per cent and that collection be taken out of the hands of the Department of Energy because the department was now tasked with what the panel described as Mission Impossible: "One cannot by definition be simultaneously responsible for both maximizing activity in the energy sector (in terms of rule-setting, licensing policy) and also ensuring that Albertans receive their fair share from energy developments...Those two mandates work in opposing directions." Given that Alberta's royalties for natural gas, oil, and heavy oil also ranked among the lowest in the world, the panel recommended a conservative 20 per cent increase over current revenues for conventional oil and gas.

Our Fair Share strongly echoed a highly critical audit of the Minerals Management Service (MMS) at the U.S. Department of the Interior. Similar to Alberta's Department of Energy, the agency is charged with collecting royalties from oil produced on federal land and Indian reservations. During a lengthy investigation into deep-sea royalties in the Gulf of Mexico, the department's inspector general, Earl E. Devaney, found a singular record of mismanagement and ethical lapses, as well as employee fears of retaliation against whistle blowers. He concluded the department didn't work for U.S. citizens any more but for oil companies.

Devaney reported that MMS auditors routinely lost files, tried to fool investigators by forging documents, and backdated documents originally reported as missing. Royalty breaks for offshore drilling were supposed to end when oil hit $34 a barrel, but the agency carelessly omitted that clause from final contracts, with the result that U.S. taxpayers lost more than $10 billion to oil companies. Devaney accused the MMS of being much too cozy with industry and said its dismal performance "demonstrates a Band Aid approach to holding together one of the federal government's largest revenue-producing operations." The inspector general has since begun investigations "of a criminal nature" into the MMS.

Our Fair Share became an immediate bestseller in Alberta. While the public supported its modest recommendations for a higher take and a new accountability regime, the oil patch responded with bitter denunciations. Taking a page from climate change scaremongers, multinational players in the tar sands threatened economic Armageddon if the province accepted the panel's recommendations. They predicted job layoffs, project cancellations, and the flight of international capital. Many outraged oil-patch executives, sounding like Mafia dons caught robbing a church, even threatened to leave the country. EnCana, which posted the biggest annual profit in Canadian corporate history in 2007 at $6.4 billion, warned that any tinkering with the royalty regimes might force the behemoth to pack up $1 billion worth of Alberta investments and move to Wyoming and Texas. (Both states, as documented by the royalty panel and the GAO, charge higher royalties than Alberta does.) Nexen, ConocoPhillips, Talisman, Canadian Natural Resources, and Petro-Canada made similar threats.

FirstEnergy Capital Corporation, a Calgary-based investment firm, sent out a newsletter hailing the province as some sort of "Albertastan" that was risking thirty thousand jobs. Another over-the-top rant came from Deutsche Bank, which said that the Alberta's royalty review panel's report read as though it had been penned by "a visiting delegation of Venezuelans." Even though Alberta's share of economic rent from hydrocarbons had declined 25 per cent over the last thirty years, the bank called the proposed increase a ruse to compensate the Albertan on the street for "self-perceived under-taxation." Wall Street energy analyst Fadel Gheit stomped and roared like Rumpelstilskin: "It's tried and true: if you really want to hurt your economy, start raising taxes (i.e., royalties) on industries that are basic to the lifeblood of your economy. It's so stupid."

Two prominent corporate leaders expressed their outrage about the panel's report in the *Globe and Mail*. Gwyn Morgan, former CEO of EnCana, described the panel's recommendations as unprincipled: "If you poll almost any society with questions like 'should the rich pay more' or 'should industry pay more,' you can count on a majority of yes answers. If you precede that poll by a government commissioned report alleging that 'the people' have not been getting their 'fair share,' the number of yes responses will be even greater." Charlie Fischer, CEO of Nexen, reflected: "When we talk about fair share, I don't know what that is...I found it really shocking to see the animosity people had for the sector."

Rex Tillerson, the CEO of ExxonMobil, the world's biggest oil company, joined the fray by arguing that Alberta should keep to the status quo: "Give us stability and honor the deal." (Unlike the U.S. convention, Alberta's oil leases plainly state that the province can raise the royalty at any time.) Tillerson did not mention that ExxonMobil has had difficulty honouring its own royalty deals. In 2001, a jury found the megafirm guilty of defrauding the state of Alabama by willfully underpaying on state royalties for natural gas wells drilled off the coast. Early in 2008, ExxonMobil agreed to pay back $300 million to the state's treasury. Four other tar sands developers—ConocoPhillips, BP, Chevron, and Shell—have also been sued by the U.S. Department of Justice for underpaying royalties. In 2000, the four firms returned nearly $300 million to taxpayers.

To most business-minded folk, the oil patch's hysterical campaign appeared as nothing more than an extreme negotiating ploy. Even the respected British energy consultancy group Wood Mackenzie admitted that "Alberta would remain one of the cheaper places to do business in the world" if the government accepted all of the panel's recommendations. Royalty experts and members of the royalty panel patiently explained that eighteen countries, including the United Kingdom, had raised their royalties in response to higher oil prices. Other commentators pointed out that a modest 20 per cent royalty increase would merely move Alberta from the bottom of the global rent pond to somewhere near the middle, slightly beneath Nigeria. Pedro van Meurs, a global expert on royalties and adviser to the blue-ribbon royalty panel, put all of the industry screaming into perspective: "In my entire 34-year career as a fiscal adviser to governments, I have never had an oil executive indicating to me that it was the right moment to increase royalties."

Evan Chrapko, a panel member, also found industry's Chicken Little act hard to stomach. Chrapko said that he was astounded that Alberta Energy did not have an accurate record of how much royalty money was owing compared to how much had actually been collected. In response to wild claims about cancelled projects and lost jobs, Chrapko asked why, "in a free country where risk-taking capitalism is supposedly the order of the day, is it up to the royalty system to single-handedly act as a magic wand against much bigger forces" such as rising labour costs? Jeff Rubin, CIBC chief economist, also had little sympathy for the hydrocarbon doomsayers: "Where are they going to go? Venezuela?"

Alberta's auditor general, Fred Dunn, weighed into the 2007 debate with more disturbing findings of gross negligence. Dunn first reported he had found no evidence of a proper royalty review. He also revealed that the province's energy department knew three years earlier that Alberta was not collecting its fair share of royalties and could have collected an additional billion dollars "without stifling industry profitability." Technical staff knew that Alberta had not been collecting its fair share since 2000 due to high energy prices. But "neither this information nor the reasons why changes have not taken place have been made public," Dunn said. The department also failed to capture another $1 billion linked to high natural gas prices. The auditor general told reporters that "the principles of transparency and accountability, I believe, were not followed." In response, Energy Minister Mel Knight refused to offer his resignation. He also denied there were any missing billions, and called Dunn's comments "a personal attack."

While the Alberta government issued one denial after another about the mismanagement of its royalty regime, Alaska came face to face with the ugly nature of royalty fraud. Similar to Alberta, the state enjoys a frontier boosterism, earns a majority of its revenue from oil and gas, and has no sales tax. But in 2005, the FBI and the IRS started to monitor the doings of Bill Allen, president of VECO Corporation, one of the state's largest oil-service companies. Over a two-year period, the FBI tracked Allen, a profane seventy-year-old, bribing Alaska's most senior and influential Republican politicians with $400,000 to keep royalties low while oil prices climbed sky high.

Investigators bugged Allen's office, capturing the whole scandal on audio- or videotape. In particular, Allen wanted to torpedo a 2006 bill before the legislature that would have raised the taxes on oil production in the state by 23.5 per cent. The FBI taped ConocoPhillips Alaska President Jim Bowles telling Allen that "if there's any way we [can] get this thing stopped, that's the best possible outcome." After the bill went down to defeat, the FBI taped Pete Kott, former Republican speaker of the Alaska House of Representatives, boasting to Allen in a Juneau hotel suite: "I had to cheat, steal, beg, borrow, and lie. Exxon's happy. BP's happy . . . I'll sell my soul to the devil."

In 2007, Allen pleaded guilty to charges of bribery, and Kott is now serving six years in jail for bribery and extortion. Alaska's big three oil companies—BP, ConocoPhillips, and Exxon—have all denied knowing anything about Allen's criminal activities. But the energy resource sector has contributed $418 million to political campaigns in the United States since 1990. Allen alone chipped in a cool $1 million.

What surprised Alaskans most about the scandal was the puniness of the bribes. Elected representatives typically sold out citizens for $2,000.

"We all thought it'd take a Mercedes or a Porsche," joked an Anchorage entertainer. "Nobody knew you could buy a politician for the cost of a used riding lawn mower." The scandal, however, did shake up the state. Alaska has a new governor, Sarah Palin, the no-nonsense mother of four children. In 2007, she restored the 25 per cent tax with a declaration Albertans haven't yet heard: "We are a state very rich in natural resources. Currently, we do not receive fair value for our resources as they're extracted and sold for us at a premium, to very hungry markets." Alaska's new Clear and Equitable Share program also closed loopholes and reduced allowances for deductions. One local politician confessed, "In a very odd sense, the FBI helped us finally fix our oil tax and finally get our fair share." Alaska now reaps 60 per cent more income from its oil than Alberta does. While $42 of every $100 oil barrel goes to Alaskan state coffers or a petroleum fund, only $26 a barrel is set aside in Alberta.

In the end, industry's temper tantrum undid the government's nerve. Premier Stelmach tabled a highly compromised royalty plan in November 2007. It denied Albertans a fair share as well as an unbiased accounting system; instead of a 20 per cent increase that would net the government an additional $2 billion a year after 2008, as the panel recommended, the province proposed a $1.4 billion increase beginning in 2010. Royalty expert Pedro van Meurs pronounced the new tar sands terms "highly detrimental to Alberta... Albertans have lost the opportunity to gain a secure and reasonable share from the rapidly increasing oil sands production." Moreover, van Meurs said, the new terms ensured that Albertans would be paying for all the cost overruns in the tar sands through lower royalties and taxes.

After the auditor general's damning report, the Alberta government tried to massage its image by hiring a well-known Tory confidant, Peter Valentine, to write a report on the government's business practices. In March 2008, Valentine dutifully reported that the Department of Energy "has designed appropriate business process and controls to support the goal of optimizing Alberta's share of resource revenue." Valentine, who served as Alberta's auditor general between 1995 and 2002, dismissed the idea of independent auditing as too expensive. He also found no conflict of interest in having the same ministry that promotes tar sands development and is directed by a former oil man collect oil royalties. Valentine, the province's former auditor general, forgot to mention that he was often reviewing his own legacy in his report. But he had to concede that staff shortages in the department and a four-year delay on tar sands audits (name an oil company that would tolerate four-year delays in basic accounting) presented a "significant risk given the critical role Audit and Compliance plays in ensuring the Department is collecting 100 per cent of what can and should be collected from industry."

Curiously, Valentine made no mention of government documents released under the Freedom of Information Act to political activist Martha Kostuch. The documents included a 2006 report that estimated the province had lost between $1.3 billion and $2.8 billion in "uncaptured economic rent" from natural gas due to outdated accounting practices. Although one team at Alberta Energy had called upon the government to "increase conventional oil and gas royalties to restore Alberta's fair share at high prices," Klein, Melchin, and then Knight repeatedly withheld that information from Albertans. Department staff had also done a royalty comparison with eight U.S. oil-producing states that showed Alberta ranking, once again, as the best oil bargain on the continent. Alberta Energy spokesman Jason Chance later admitted to the *Edmonton Journal* that the government released the uncensored material to Kostuch by mistake.

In May 2008, Auditor General Fred Dunn repeated his charge that the Alberta government had failed to collect billions in royalties since 2001. Because Stelmach's new royalty regime makes no allowance for $130-a-barrel oil, Pedro van Meurs warned that Alberta still lags behind most countries in responsible oil-wealth management. "They are not capturing the proper rent. You leave a bundle on the table. It is just unbelievable," he said. The province said it would make no adjustments.

Alberta's refusal to collect, let alone audit, its fair share of royalties is matched only by its refusal to save for a rainy day. In 1976, Premier Peter Lougheed set up one of the world's first sovereign petroleum funds, the Heritage Fund, and dedicated 30 per cent of all non-renewable resource revenue to it. Later governments stopped contributing money to the fund in 1987. Under Klein, the fund languished as a poorly managed slush fund the government often looted at its own discretion. Its value today stands at a measly $17 billion. Norway took Lougheed's brilliant idea, surrounded it with stricter rules, and mandated that 94 per cent of all oil revenue go directly to its fund. Since 1990, Norway's well-managed fund has amassed $400 billion in savings for the future. Both the Alberta Chamber of Commerce and the OECD have decried Alberta's ad hoc approach to savings, calling for a renewal of the Heritage Fund with firm goals and accounting rules. But Energy Minister Mel Knight says Alberta has nothing to learn from Norway.

In June 2008, Jim Roy, a former senior analyst with Alberta Energy, took a fresh look at Alberta's royalty review and the government response. Klein had fired the former civil servant and several fellow analysts in 1993 because they believed the government should calculate royalties in the public interest. Klein favoured a monologue with industry.

Roy, who now advises nations around the world on royalty issues, described many of the panel's proposals as "insufficient and unworkable." Moreover, he concluded, government amendments to the panel's proposals meant that the citizens of Alberta will actually be making less money than they did prior to the review. "It's a reduction, not an increase," he said. "Alberta is still at the bottom of the list. Our royalties are less than Saskatchewan and British Columbia." What Alberta has, added Roy, is a third-rate royalty target geared for "rapid development."

In a recent article in the *New York Times Magazine,* reporter Tina Rosenberg noted that "finding a hole in the ground that spouts money can be one of the worst things to happen to a nation." She reviewed the dismal history of Venezuela and concluded that the best thing an oil-rich jurisdiction can do "is to keep the oil private, watch it carefully and tax the hell out of it." Better yet, raise the royalties and have companies compete with one another in open bidding for access to the oil. Rosenberg concluded that nationalizing a resource is no substitute for good oversight, accountability, and transparent, hands-off management of royalties. Alberta, a bona fide petrostate, has repeatedly failed on all accounts.

TWELVE

THE FIRST LAW OF PETROPOLITICS

....

"Control oil and you control nations; control food
and you control the people."

HENRY KISSINGER, U.S. NATIONAL SECURITY ADVISOR, 1970

OIL HAS FANTASTIC powers, and like the genie from *Arabian Nights*, it
can grant political wishes both fair and foul. This is why U.S. oil baron
John D. Rockefeller, in a moment of reflection, called oil "the devil's
tears," and why Sheikh Ahmed Zaki Yamani, in a moment of exasperation, wished that Saudi Arabia had discovered water, and why the
Venezuelan writer José Ignacio Cabrujas wrote mischievously that oil
can create "a culture of miracles" that erases memory.

The First Law of Petropolitics is not complicated. You won't hear it
discussed at Calgary's Petroleum Club or in Ottawa's corridors, because
the obvious rarely makes idle talk among the powerful. But the law does
explain the bizarre and unsettled state of Canadian politics, our obses-
sion with North American union, the authoritarian character of the
Alberta government, the impervious nature of the Stephen Harper
regime, the nation's dismal climate change record, and the incredibly
rapid development of the tar sands. It also explains why the world's
largest, dirtiest energy project has become the dominant driver of

Canadian and North American economic life without so much as a debate in Canada's House of Commons or the U.S. Congress.

New York Times columnist Thomas Friedman unveiled the law in a 2006 issue of *Foreign Policy Review,* and it goes like this: the price of oil and the quality of freedom invariably travel in opposite directions. As the price of crude oil climbs higher in an oil-dominated country, poor or rich, secular or Muslim, that country's citizens will, over time, experience less free speech, declining freedom of the press, and a steady erosion of the rule of law. Neither Texans nor Canadians are exempt. Friedman calls it "the axiom of our age."

Friedman argued that the First Law explained the emerging petro-tyrannies of Venezuela, Iran, Nigeria, and Russia. When oil hovered around $25 a barrel, Nigeria politely offered to investigate human rights abuses and root out corruption. Iran talked about dialogue and peace. But as soon as oil roared past $60 a barrel, these noble intentions evaporated. When oil sold cheaply, Russian President Vladimir Putin behaved like an enlightened political reformer. During his last years as president, he acted more like an eighteenth-century czar; he systematically used the nation's oil and gas resources to boost control of the energy sector, blackmail other nations, buy out newspapers, silence journalists, and generally entrench authoritarian rule.

Most academics know the First Law of Petropolitics as "the resource curse" or "the paradox of plenty." For years, scholars have noted that oil-rich states rarely achieve political maturity or economic diversity and inconsistently share the resulting wealth with their citizens. A resource boom can single-handedly hollow out an economy and sicken the nation with Dutch Disease: in the 1970s, natural gas discoveries so inflated the value of Dutch currency that the resource nearly killed Holland's manufacturing base. Oil can reduce every other economic sector, no matter how tall, to a midget.

Middle East specialists have long suspected that the curse explains the abiding dearth of democracy in that region's oil-rich kingdoms. Michael Ross, a soft-spoken political scientist from the University of California, checked out the idea in 2000 and proved the specialists right. Ross examined a number of social and political measurements, such as taxes and military spending, from 113 different states between 1971 and 1997 and found that a "single standard deviation rise" in oil wealth directly corresponded with a 0.72 drop on a democracy scale. The curse was very much alive.

Ross identified three subtle ways that oil hinders democracy. The first is the taxation effect. Governments with lots of oil revenue don't need to tax their citizens to govern. All they have to do is approve another tar sands project, license another gas well, or put more land up for sale. The first thing most newly minted petrostates do is reduce or eliminate taxes. Most of the U.S. Gulf states, for example, don't have any taxes. Nor does Wyoming, a treasure chest of natural gas and coal.

In the absence of taxes, people are less inclined to be vigilant about how their government spends money, and they are less inclined to ask questions. In many jurisdictions, such as Alberta, they may not even bother to vote. The province has one of the lowest voter turnouts in North America. Oil-stoked governments, in turn, are less inclined to listen to their citizens or to represent their concerns. When governments collect more revenue from hydrocarbons than they do from taxpayers, they eventually forget whom they serve. "I think this explains why even relatively democratic countries see less accountability in their government," says Ross. Thomas Friedman says that while the motto of the American Revolution was "no taxation without representation," the credo of "the petrolist authoritarian is no representation without taxation."

Second, oil-addled governments often spend their petrocash on patronage or state-funded programs that discourage thought, debate, or dissent. Throughout the Middle East, governments have deliberately dismantled independent civil groups while creating their own multistakeholder associations. In both Mexico and Indonesia, oil has consistently propped up one-party rule. Third, according to Ross, oil wealth gives wayward governments the means to invest heavily in guns, tanks, and "the apparatus of repression." When tax breaks and an orgy of patronage fail to buy people's allegiance, oil-rich states just call in security.

Ross recently dug deeper to find that oil changed the electoral fate of governments in two amazing ways. "The more oil and gas a government has access to, the wider margins it won in elections and the longer its leader stayed in power," he concluded. In other words, oil gives governments, whether ruled by kings or republicans, the financial ability to buy votes or influence the political marketplace.

Authoritarian oil-based regimes just don't decorate the jungles of South America or the deserts of the Middle East. They dot the landscape of North America, home to the world's first oil discoveries. Two U.S. political scientists, Erik Wibbels and Ellis Goldberg, recently asked if the resource curse had influenced the development of the United States. Sure enough, they found it had.

In the 1930s, Texas, California, Louisiana, and Oklahoma stood out as the world's major oil producers. In these states, oil wealth performed its usual magic: it powered political machines and fed rampant corruption as well as helping to build schools. According to Wibbels and Goldberg, oil also reworked electoral patterns. Oil-gushing states typically recorded a much higher gap in the number of votes between winners and losers (incumbents typically captured an 8 per cent higher share of the vote) wherever the government's dependence on oil revenue totalled 20 per cent or more. The parties of the winners, of course, tended to tax less at the same time as citizens witnessed a serious decline in the integrity and quality of civil institutions.

In a separate 2008 study, Wibbels and Goldberg analyzed electoral data spanning seventy-three years in the United States. They again found that oil, gas, and coal had left a recognizable stain on the democratic cloth. The electoral record indicated that "politicians in resource rich states have shown considerable skill in using mineral wealth to their advantage." Oil consistently allowed those politicians to buy public support and enrich their friends, thereby stunting the development of a viable opposition and of related democratic institutions. Oil also insulated bad government by giving it the capacity to survive public disapproval with lots of cash. For every 1 per cent increase in resource dependence, an oil-rich state usually upped its per capita spending by

$3.43. The authors concluded that "political incumbents in resource abundant polities with fair and free elections manage to win by larger margins and preserve vote shares in the face of adverse circumstances in a way that politicians without access to mineral rents will not."

Huey Long, the populist demagogue of Louisiana, made a tidy example of how the First Law of Petropolitics fuels authoritarian regimes. The governor came to power under the slogan "Every man a king," and he ruled Louisiana in the late 1920s much like a monarch. Although Long used oil wealth to build schools and improve public health, "the Kingfish" also used the money to fashion a political machine that, as Wibbels and Goldberg noted, "more nearly matched the power of a South American dictator." The machine behaved much like Louisiana's previous unelected ruler, the Standard Oil Company. Long's political network took kickbacks, exported oil illegally, and boosted the profits of oil companies in which Long supporters held stock.

Texas, the capital of oil for the western world, has long saluted the First Law as a distinct petrostate. It has even sent two oil men from the same powerful family to the White House. Financed by Big Oil, both presidents have acted as shameless advocates for the industry. In Texas, the resource has created such "an equilibrium of interests between industry and politics" that George Bush Jr. has no problem holding hands with Saudi princes. Even seasoned Republicans admit, as the *Observer* reported in 2002, that Texas has "vending machine politics: you puts your money in and you gets your product out."

By any conservative definition, Alberta makes an attractive poster child for the First Law. Oil and gas revenues compose a quarter of the province's GDP and provide the government with more than 30 per cent of its total revenue. Not surprisingly, the province has been ruled by the same political party for thirty-eight years. Like most Middle East countries, Alberta has no sales tax. It also has the lowest overall taxes in Canada, with no general capital or payroll taxes.

Since the discovery of oil and gas in the 1920s, the province's politics have faithfully mimicked those of most petrostates. Ruling parties typically win by large margins, and their leaders stay in power much

longer than in any other jurisdiction in Canada. Although naive commentators call Alberta a political "maverick," it is nothing of the sort.
Oil and gas wealth have merely bent its political character, leaving it
fat and lazy.

As in oil-rich Louisiana, Albertans commonly call their political
leaders kings. Ralph Klein, a boozy journalist, gambler, and free-market
version of Huey Long, ruled the province for fourteen years with massive pluralities. The media affectionately dubbed him King Ralph. As
soon as Klein's successor Ed Stelmach won massive pluralities in 2008,
the media obediently crowned him King Eddy. In bitumen-soaked
Alberta, even journalists forget an elementary school lesson: kings do
not rule democracies.

Citizen engagement is largely a spent force in Alberta, even compared to petrostates such as Venezuela. In each subsequent election,
fewer citizens bother to vote. Only 40 per cent of the electorate marched
to the polls in the 2008 provincial election, the lowest voter turnout in
the history of Canada. This dismal pattern worries thoughtful Tories
such as well-known blogger Ken Chapman: "If we do not start to have
politics that are relevant and engaging to our citizens we open our-
selves up to all kinds of problems from corruption and demagoguery to
despair with a disintegration of our sense of social cohesion and common purpose."

Every petrostate develops its own unique authoritarian style. Some,
such as Venezuela, use the money to insert the state into places it does
not belong. Others, such as Alberta, neglect to collect the money and
allow the marketplace to govern in places where it should have no
authority. Klein and his successors have hijacked the machinery of the
state to unduly enrich multinational corporations.

King Ralph behaved like a Huey Long in reverse, a Robin Hood for
the rich. He started by undoing all the democratic controls on petro-
wealth that Premier Peter Lougheed had put in place in the 1970s. To
minimize the resource curse, Lougheed established a Norwegian-type
regime long before Norway improved on his ideas. He increased royalties to 40 per cent of total oil and gas income and set up the Heritage

Fund for the future. He also established a Crown corporation, the Alberta Energy Company, that Albertans could invest in. The company gave the province's citizens an open window into the oil patch. But Klein undid the whole works. In 1993, he fired a host of economic analysts in Alberta Energy because, as one former civil servant recalled, "He wanted industry to tell him what to do." Klein let royalties drop to 15 per cent of the hydrocarbon pie, which made the province one of the most enriching regimes anywhere for multinationals. Instead of saving for peak oil, such as Alaska and Norway do, Klein capped contributions to the Heritage Fund, and it stopped growing altogether. In 1996, he sold off the Alberta Energy Company, along with some of the province's richest hydrocarbon assets, at a third of their market value, to the company that became EnCana.

Klein also used every petrofuelled machination documented by the political scientists to buy the fidelity of the electorate. When a botched electricity deregulation plan drove electrical prices skyward, the king dipped into his handy hydrocarbon revenues (largely from natural gas sales) and spent $4 billion on power and natural gas rebates before the 2001 election. In the same situation, most other governments would have gone bankrupt or suffered defeat at the polls. Klein just bought another political victory.

King Ralph, who openly admitted that he preferred governing "on auto pilot," vowed that taxes would only go down. True to his word, he used his hydrocarbon revenue to lower income and corporate taxes in the province. He even handed out $400 prosperity cheques, or "Ralph bucks," to the electorate in 2004 at a cost of $1.4 billion. It's no accident that Kevin Taft, the lacklustre leader of Alberta's fledgling Liberal Party, called his book about Canada's hydrocarbon kingdom *Democracy Derailed.*

The derailing has taken many forms besides the passing out of "Ralph bucks." Distinctions between the business of hydrocarbons and civic affairs, for example, have all but disappeared in Alberta. Within six months of quitting his job as Alberta's number-one petrobully, Klein became a paid senior business adviser in the oil patch for Borden Ladner

Gervais LLP. He told the *Star Phoenix* in 2008 that he now promotes multinationals and their tar sands developments only "if they pay me." Klein also writes reports for conservative think tanks that advocate the laissez-faire program he promoted as premier: bargain-basement-priced hydrocarbons and a "long-term continental strategic framework" that supports further integration of North American energy markets. His former chief of staff, Peter Elzinga, leapt from Klein's office to the employ of the tar sands giant Suncor as a lobbyist in 2004, only to jump back into politics as the executive director of Alberta's Conservatives nine months later. Three months after quitting politics, former Alberta Energy Minister Greg Melchin joined the board of an oil company, while the former minister of economic development, Mark Norris, was appointed chairman of Wescorp Energy Inc.

Alberta's former ambassador to the United States, Murray Smith, now sits on TD Bank's Energy Advisory Board, but he gives the same speeches he gave as a provincial energy minister. In 2007, Premier Ed Stelmach hired Heather Kennedy, a Suncor vice-president, to direct the Oil Sands Sustainable Development Secretariat. Her job is to help sort out the chaos caused by rapid tar sands development. The highly competent oil-patch executive will serve as an assistant deputy minister in the provincial treasury department but be paid by her company. It's a unique relationship.

Given their one-dimensional character, oil regimes generally fear transparency, and Alberta is no exception. The province has one of the most secretive governments in Canada. In 2006, Alberta's Conservative government made it legal for its petropoliticians to lock away internal audits for fifteen years and for government ministers to keep their briefing binders out of public view for five years. Freedom-of-information requests take months and cost a small fortune to obtain. Most material arrives blacked out.

Critical information has a way of disappearing in a petrostate. When a confidential 2006 report by a team of anti-terrorism experts documented "serious concerns" about the state of security at the world's largest energy project, Klein refused to release the document. The report warned that "an attack against any of the oilsands facilities could be easily achieved" and said that the tailings ponds seemed particularly vulnerable: "If the berm [of Syncrude's dam] was breached, the ensuing environmental impact would not only close down the oilsands, it would cause long-term damage to the eco-structure of the Athabasca River." According to Nathan Jacobson, a Toronto businessman who is one of the report's authors, the document was deep-sixed. The security team also found a sophisticated bugging device in the office of the Treasury Department in the Terrace Building at the Alberta Legislature. The public was never told that a foreign government or corporation probably knew about the contents of the province's budgets before Albertans did. A 2007 report by the Canadian Security Intelligence Service, released under the Freedom of Information Act, also concluded that the tar sands industry represents an "ideologically attractive and strategic target" for groups like Osama bin Laden's al-Qaida. But in Alberta, the alarming political risks of becoming the world's number-one oil supplier to the United States is never discussed.

Petrostates also know how to control the conversations of ordinary people. The Alberta government currently spends $14 million a year and employs 117 full-time staff in its Public Affairs Bureau to tell Albertans what to think. It has devoted another $25 million to convincing both Alberta's citizens and U.S. oil consumers that the tar sands are greener than Kermit the Frog. The Public Affairs Bureau works much like the Politburo in the former Soviet Union. Not even George W. Bush Jr. has employed a propaganda arm this large in the White House.

The tone of the Alberta government has become increasingly authoritarian. Premier Ed Stelmach declares that he can't "touch the brakes" on rapid development in the tar sands, any more than his counterparts in Venezuela or Russia can, say, touch the brakes on aggressive nationalization. Yet only drunks and hit-and-run drivers use this sort of language without irony. Stelmach, now Canada's highest-paid premier, has also begun to call opposition parties "subversive."

Although Alberta has many strong environmental rules, it rarely implements them. A recent Cornell University doctoral study on the province's resource curse concluded that "responsibility buck passing" and lack of public input, combined with no cumulative environmental studies and a steady "institutionalized development bias," have made the province's environmental department toothless. The government has instructed Alberta Environment employees, for example, to refer to air pollution as "air emissions." Alberta's environment minister, Rob Renner, talks like a minister of development. Renner disclosed in 2007 that he wasn't concerned about the hectic pace of oil and gas activity: "The speed with which economic development takes place is not something the government has control over... slowing the pace inevitably results in stopping the development and it's difficult to get it going again." The minister also confirmed that "it's not the role of Alberta Environment to advocate on behalf of the environment." (In 1974, Alberta's first environment minister, William Yurko, said exactly the same thing.) To date, the department has been largely a silent bystander in the tar sands. A recent analysis of Alberta Environment's quarterly reports revealed that most tar sands projects, despite leaks, spills, and upsets, faced only a single fine between 2006 and 2007. With regard to water quality, the federal government's enforcement of the Fisheries Act between 1988 and 2005 was equally uneventful.

In recent years, Alberta has increasingly sacrificed the rule of law to ease the flow of energy exports. Whenever open public debate threatens to challenge a government-sanctioned energy project, the Energy Resources Conservation Board, a de facto rubber stamp for oil and gas development (it approves more than 94 per cent of all applications), shuts down public participation, citing "security" reasons. In 2007, the board even hired spies, at a cost of $100,000, to gather "covert intelligence" on rural landowners peacefully questioning the pace of energy development in their backyards. Premier Stelmach initially defended the spying. In a petrostate, even the voice of a disenfranchised senior citizen can be perceived as a dangerous threat.

Elected bodies no longer pull much weight in Alberta, either. In 2007, the council of the Regional Municipality of Wood Buffalo, a democratically elected body representing the hard-working citizens of

Fort McMurray, presented compelling arguments for a slowdown of tar sands development in order to preserve some sense of community. The ERCB, a government-appointed body with no public oversight, overruled the municipality every time. Not surprisingly, only 21 per cent of the people in the Fort McMurray area voted in the 2008 provincial election. They know that bitumen calls the shots, and many of them won't be staying long.

The democratic gap between the rulers and the ruled grows wider every day. Polls show that Albertans overwhelmingly favour real reductions in carbon emissions, yet their government champions a laughable program to reduce emissions by 14 per cent by 2050. Most people want a slowdown in the tar sands, but the government will hear nothing of it. Rural Albertans ask for tough groundwater protection but get more oil and gas drilling in their backyards instead.

Exercising freedom of expression in Alberta can be dangerous, as Dr. John O'Connor found when he called publicly for a health study of communities downstream from the tar sands. Following O'Connor's political persecution, the Canadian Medical Association passed a motion in 2007 urging that doctors be protected from "reprisal and retaliation" when they serve as community advocates.

Rapid development of the tar sands is transforming the Canadian government into a petrostate, too. Given that Canada now produces more oil than Kuwait, that it derives nearly 9 per cent of its gross domestic product from energy exports, and that it will soon be the globe's fourth-largest exporter of oil, the Conservatives in power have increasingly saluted the First Law of Petropolitics.

Stephen Harper, Canada's own blue-eyed sheik, has become an able spokesman for bitumen. (Harper's home is in Alberta, where the largest bitumen producer, oddly enough, is Imperial Oil, which once employed his father.) Harper's best friends include a bevy of climate change deniers such as petroleum geologist John Weissenberger and University of Calgary academic Barry Cooper. These connections partly explain why the prime minister dismissed Canada's international obligations under Kyoto to reduce carbon emissions with all the flair of a Hugo

Chavez violating legal agreements with multinational companies. To successfully stall any real action on energy or carbon conservation, Harper appointed Rona Ambrose, the daughter of an oil executive, as his first environment minister.

Harper's oil allegiances are well known. He tried to appoint Gwyn Morgan, a Tory fundraiser and former head of EnCana (North America's largest natural gas producer and one of the largest holders of tar sands leases), to oversee government accountability. Morgan, a Canadian version of Dick Cheney, proposed to work for free. The position would have made Morgan, a champion of continental integration, the key overseer of many Crown corporations, including the National Energy Board. Much to Harper's dismay, parliamentarians rejected the appointment. Al Gore was right when he observed that "the financial interests behind the tar sands project poured a lot of money and support behind an ultra-conservative leader in order to win the election and to protect their interests."

Since his 2006 election, Harper has steadfastly earned a reputation as a secretive and heavy-handed leader. The country has no deputy prime minister, and cabinet ministers rarely speak out of line. To ask questions, journalists must scurry to get on a preapproved list, cap in hand. "To search the annals for another Canadian PM who accumulated so much cold-blooded authority in such a short time is to come up empty," wrote *Globe and Mail* columnist Lawrence Martin. That's saying a lot. Canada, a country founded on the exploitation of one staple after another, from furs to uranium, has a long tradition of caudillo-like leaders.

Bitumen has contaminated the fiscal machinery of Harper's government too. Every day, the feds give the tar sands industry a million dollars' worth of tax breaks. It took a formal petition filed with Canada's auditor general by the church group Canadian Ecumenical Justice Initiatives to force the government to confirm the scale of the giveaway. Harper's government, however, refused to answer the petition's key question: "Why does Canada spend millions of dollars on subsidizing oil and gas industries—a prime cause of climate change—and so little money on great alternatives?"

Bitumen has also begun to reorient the federal bureaucracy. In 2004, the National Energy Board (which some critics suggest should be renamed No Energy Policy), signed a Memorandum of Understanding with the U.S. Federal Energy Regulatory Commission (FERC) to expedite "coordinated action on significant energy infrastructure projects." Similar memorandums have been signed to expedite bitumen pipelines. In 2008, the militaries of Canada and the United States co-signed a Civil Assistance Plan that allows soldiers from either country to curb civil unrest, defend oil facilities, or "support rapid decision-making in a collaborative environment."

Foreign Affairs, when not issuing press releases on Canada's role as a northern Saudi Arabia, operates a new "energy secretariat." The bitumen-friendly agency says that the government must resist "efforts to label one form of energy as appropriate such as renewables and others as inappropriate such as hydrocarbons and nuclear." Natural Resources has a new Energy Infrastructure Protection Division solely concerned with the protection of critical pipelines and refineries. The division also participates in Security and Prosperity Partnership initiatives, such as the North American Energy Working Group, that publish reports on how "the oil sands can make a truly significant contribution to North America's energy supply and security." Canada is increasingly a country about bitumen, for bitumen, and by bitumen.

In 2006, the Library of Parliament released a little-read report entitled *Energy Resources: Boon or Curse for the Canadian Economy?* that found increasing evidence of the resource curse. The report concluded that Canada "does appear to have some symptoms of the Dutch Disease as can be seen in the relatively high value of the Canadian dollar and manufacturing job losses." It recommended that Canada follow the example of Norway and abide by the protocols of the International Monetary Fund, which advises oil-rich nations to separate oil income from other revenues and set up a dedicated resource fund.

To date, the federal government has ignored this advice. In 2006, Canadian governments garnered $26 billion in royalties, lease bids, and income taxes from oil and gas projects. Of that sum, Ottawa pocketed about $5 billion in corporate income taxes from the tar sands. By 2020, the federal government will have made at least $50 billion from rapid tar sands development. True to the First Law of Petropolitics, government has used this windfall so far to reduce corporate taxes and slash 2 per cent off the federal sales tax. While Norway has kept the resource curse largely at bay with clear accounting and its dedicated oil/pension fund, Ottawa has spent the cash to win friends and influence elections.

The increasingly tyrannical nature of bitumen and its public servant, the federal government, openly revealed itself at an unusual hearing of the Standing Committee on International Trade in May 2007. The committee was studying the benefits of the Security and Prosperity Partnership, which advocates for a North American economic union and total energy integration.

Gordon Laxer, an outspoken nationalist and director of the Edmonton-based Parkland Institute, made the mistake of raising a number of very conservative arguments at the hearing. First, he accurately reported that Canada had no energy plan. While Canada now exported most of its oil to the United States, half the country (including Quebec and the Atlantic provinces, whose workers toil in the sands) remained dependent on oil imports from Algeria, Saudi Arabia, and Iraq. "How secure is that?" Laxer asked. He also wondered why the National Energy Board had conducted no studies on security of supply and why Canada, unlike most developed countries, kept no oil in strategic petroleum storage for emergencies.

Committee chair Leon Benoit, a Member of Parliament from Alberta, intervened to tell Laxer that he was off topic. Drawing upon instructions contained in a two-hundred-page petrostate manual on how to control government committee meetings and gag dissent, Benoit tried to bully Laxer into silence.

"Mr. Laxer, if you are here to discuss the energy security of Canadians, then you are off topic of the study."

"I don't see that."

"We are here specifically to talk about the Security and Prosperity Partnership of North America."

"Isn't it [Canada] part of North America?"

"Mr. Laxer, please wait until I'm finished."

"I'm sorry."

"If you're here to talk about energy security as a general topic, without making that connection, then you're off topic for today."

When Laxer continued to highlight more facts about the adolescent state of energy policy in Canada, Benoit cut him off and stormed out of the room, as advised by the Harper manual. His temper tantrum illustrated the long shadow of the First Law and proved that "emerging energy superpowers" have little tolerance for the inconvenient debates that keep democracies democratic.

The resource curse has invaded the North, once strong and free. In the absence of proper safeguards and transparency, hydrocarbons and democracy mix no better here than they do in Nigeria, Russia, or Texas. Easy wealth has turned Alberta into a petrotyranny, while Canada has adopted all the trappings of an impervious oil kingdom, with a profound bitumen bias. As Canadian political leaders behave and talk more and more like careless Saudi princes, the devil's tears fall in one endless stream. Oil corrupts and corrupts absolutely.

Thomas Friedman offered but one antidote to the rising price of oil and its authoritarian proclivities: "Thinking about how to alter our energy consumption patterns to bring down the price of oil is no longer simply a hobby for high minded environmentalists or some personal virtue. It is now a national security imperative."

To put it plainly, citizens of Canada and the United States who value democracy at home and abroad must consume less oil.

THIRTEEN

EIGHTH WONDER OF THE WORLD

. . . .

"It is as though Mad Hatters and March Hares are in charge of
recent energy policy everywhere in North America."
THE RIGHT HONOURABLE EDWARD SCHREYER, FORMER MANITOBA
PREMIER AND GOVERNOR GENERAL OF CANADA, 2004

ALBERTA POLITICIANS AND many oil executives tout the tar sands as a
global energy lifesaver. While the *Wall Street Journal* points to the tar
sands as "stark evidence that the world isn't about to run out of oil," the
Paris-based International Energy Agency counts on rapid tar sands
development to "relax oil markets" and make the world a safer place.
The website of engineering giant Klohn Crippen Berger promises that
Canada's national treasure "could potentially satisfy the world's demand
for petroleum over the next century." Economists seem to cheerfully
agree that the tar sands will pump a trillion dollars into the economy by
2020. They also predict that the sands will forge a new oil and energy
service nation. "Energy is important to Canada because it sustains
manufacturing jobs in Ontario, provides employment in Newfoundland and is a source of significant revenues that sustain our lifestyle
across Canada," the director of the Alberta Energy Research Institute,
Dr. Eddy Isaacs, recently told the Canadian Parliament. "Canada has

the potential to become the world's energy provider." The influential oil-patch newsletter *First Commentary* argues that the tar sands will allow "Canada to take her rightful role as a release valve for geopolitical uncertainty." Paul Michael Wihbey, president of the Washington, D.C.–based Global Water & Energy Strategy Team, and an advisor to both the Alberta and Saskatchewan governments, asserts that the tar sands, combined with unconventional natural gas in the Rocky Mountains, will "become the most important axis of the global economy of the twenty-first century." With the globe's population growing and our energy consumption climbing, the Alberta government asks "where does the world get its energy from?" only to answer "one of the world's largest deposits of oil."

But none of this is true. Oil is the lifeblood of modern civilization, and the world has become a bleeding hemophiliac. Chinese factories, American cornfields, and Saudi air conditioners have pushed world oil demand to 85 million barrels a day, and the tar sands represent only 1.5 per cent of that. Even if production were to reach the vaunted five million barrels a day, the tar sands would barely supply 5 per cent of the world's oil consumption, a drop in the global bucket.

More than 50 per cent of the world's oil comes from massive oil fields, the so-called supergiants, all of which are declining. Peak oil explains why almost every major multinational and state-owned oil company has put up a shingle in Canada's Great Reserve. The tar sands are the last place in the world where oil companies can make an investment and grow production. Peak oil also explains why most of the world's largest importers of crude oil have a presence in the tar sands too. "In the big picture, deepwater oil and the oilsands are the only game left in town," notes CIBC chief economist Jeffrey Rubin. "You know you are at the bottom of the ninth when you have to schlep a tonne of sand to get a barrel of oil." Even King Abdullah of Saudi Arabia has said, "The oil boom is over and will not return. All of us must get used to a different lifestyle."

Kjell Aleklett, director of the Uppsala Hydrocarbon Depletion Study Group, told the U.S. House Subcommittee on Energy and Air

Quality in 2005 that fifty years ago, the world annually burned up four billion barrels of oil and discovered thirty billion more. "Today we consume 30 billion barrels per year and the discovery rate is dropping toward 4 billion barrels per year," he reported. Since 1900, no country has experienced an increase in its GDP without an increase in its use of oil, as China and India are now grandly demonstrating. "Animals that face food shortages have a hard time adjusting and usually their populations decline. Some believe that we as human beings will face a similar situation," Aleklett said.

After U.S. energy analyst Robert Hirsch lamented in 2006 that the world has never faced a problem as comprehensive as the end of cheap oil, Aleklett and two graduate students put together their own Herman Kahn-like crash scenario for the tar sands. The best their megaproject could muster in the short term was 3.6 million barrels a day by 2018. Even to reach that target, Canada would have to choose between exporting natural gas to the United States or burning most of its reserves in the tar sands to melt bitumen. No further growth in production could take place without massive investments in nuclear power to generate steam for in situ production. "While the theoretical future oil supply from the oil sands is huge, the potential ability for the Canadian oil sands industry to meet a growing world oil demand is not based on reality," Aleklett and his co-authors found. Rising tar sands production won't compensate for falling production from the North Sea, let alone for the collapse of Canada's own conventional production. The tar sands can't prevent peak oil, and Aleklett recommended in an e-mail that "Canadians should try to make the development [there] as environmentally sound as possible."

In 2007, the Energy Watch Group in Germany offered a similar assessment. It noted that seventeen of the world's largest oil companies (most of whom are tar sands investors) have failed to increase production in the last ten years despite higher prices and innovative technologies. The group predicted that global oil production will begin to decline so rapidly that by 2020 it may be impossible to close the gap with unconventional sources such as the tar sands: "Things might happen which we have never experienced before and which we may never experience again." As did Aleklett, the Germans found that the natural gas shortage, greenhouse gas emissions, and water scarcity will limit tar sands production: "It is not likely that unconventional oil sources in Canada will compensate for the future decline in worldwide conventional oil production." Their report also suggested that the automobile industry "might perceive higher greenhouse gas emissions of fuels from non-conventional oil sources as a nightmare."

Houston investment banker Matthew Simmons is even more blunt. His book *Twilight in the Desert* exposed the dwindling state of Saudi Arabia's reserves, and like an increasing number of analysts, Simmons says switching to low-quality bitumen from conventional crude is like trading a Mercedes for a beat-up jalopy. "If I were a Canadian, I'd make it illegal to use precious natural gas and potable freshwater to turn gold into lead in the tar sands," he says. His recommendations for policymakers are direct: go slowly, charge for water, cap tar sands production, and "find some other way to produce this atrocious resource other than using scarce natural gas . . . To get more addicted to the tar sands doesn't make any sense to me." Simmons doesn't think the tar sands megaproject can reach three million barrels without destroying Alberta.

Dave Hughes, Canada's leading peak-energy analyst, also predicts that tar sands production cannot deliver much more than three million barrels a day without enormous environmental and resource sacrifices. He calculates that it would cost an additional $100 billion to grow production to four million barrels a day and wonders if that is a desirable investment. "The oil sands should be viewed as a marginal interim supply that serves as a bridge to prepare for a less energy-intensive future," he says.

As Herman Kahn predicted, the tar sands have become a global energy playground. The governments of Alberta and Canada act like joyous peanut hawkers who can't believe the size of the crowd. After making the province one of the world's most generous fiscal regimes for oil, Alberta has simply pretended that it no longer owns the resource, let alone has the ability to control the pace of its liquidation. And if the marketplace is allowed to rule, the frenzy has just begun.

"History is likely to write scenarios that most observers would find implausible," Kahn said. Right now, Canada is implausibly digging up and replumbing an area the size of Nepal, not to save the world or to ensure its own energy security but to keep wealthy oil companies in business and to supply a fading empire with dirty oil. To date, not one national newspaper has bothered to assign a reporter to the Athabasca region to daily cover this nation-changing event. The country seems unmoved by the political implications of rapid energy integration as well as by the moral consequences of converting a forest into a carbon storm and the planet's third-largest watershed into a petroleum garbage dump. But the money is flowing, and we can celebrate our status as the world's first energy superpower without an energy plan. No wonder Premier Ralph Klein once lovingly called the tar sands "the eighth wonder of the world."

FOURTEEN

TAR AGE AHEAD

....

"Our principal impediments at present are neither lack of energy or
material resources nor of essential physical and biological knowledge.
Our principal constraints are cultural. During the last two centuries we
have known nothing but exponential growth and in parallel we have
evolved what amounts to an exponential growth culture, a culture so
heavily dependent upon the continuance of exponential growth for its
stability that it is incapable of reckoning with problems of non growth."
MARION KING HUBBERT, SHELL GEOPHYSICIST, 1976

THE QUARRY OF the Ancestors, about forty-five miles north of Fort
McMurray, lies in a piece of bush off the Highway to Hell, on the other
side of what workers once called "the bridge to nowhere." Now the
bridge delivers thousands of workers to Shell's Albian Mine, Imperial's
Kearl Mine, and a property owned by the Chinese-run outfit Syneco.
Near the river a small patch of the forest has been spared the usual
truck-and-shovel makeover. In 2006, archeologists found more than
300,000 artifacts on the site, including knives, scrapers, stone flakes,
tiny microblades, and even a spear point with ten-thousand-year-old
mammoth blood on it. The stories that the Dene told Charles Mair
about "a monster many times larger than a bison" were true.

Calgary engineers Derrick Kershaw and Donald Dabbs, senior vice-presidents of Birch Mountain Resources, work nearby. Their company is building the largest limestone quarry in Canada right next door to the site. (Rapid tar sands development isn't possible without tons of aggregate for roads and building.) Birch Mountain has generously set aside 20 per cent of its reserves to protect this important landmark.

At the bottom of a shallow hole in the forest, surrounded by tamarack and spruce, sits a smooth boulder made of Muskeg River microquartzite. It doesn't look like much. But Dabbs and Kershaw tell a fantastic geological story about it. Ten thousand years ago, a catastrophic climate change event created this quarry. A great glacier rapidly melted and released a massive amount of water, which scraped away the forest floor and exposed a rare, fine-grained sandstone. The stone made such fine tools that an ancient tribe of entrepreneurs set up a seasonal summer camp here. The stone makers eventually produced weapons for most of Western Canada.

Says Kershaw, a cheerful, tall engineer with a British accent: "The quarry was a source of projectile points and was a hell of a lot more valuable than oil. It fed families. These people were kings in their day. It was their boom. It was their currency, just as we are kings today with oil. So the cycle continues."

The original footprint of the quarry of the ancestors probably occupied a square mile. By contrast, the tar sands are well on their way to consuming a piece of real estate as large as Belgium. Kershaw doesn't think the people of Fort McMurray have wrapped their minds around the fact "that Saudi Arabia is coming to northeast Alberta."

DAVE HUGHES GIVES talks across the country about the finite nature of oil quarries. Hughes explains that in 1850, 90 per cent of the world travelled by horseback and heated with renewable fuels. Today, 89 per cent of the world is dependent on hydrocarbons and cheap airplane flights. The world now consumes forty-three times as much energy with seven times the population. "What are the sustainable implications of that?" Hughes asks.

Burning the energy equivalent of a barrel of oil to retrieve four barrels of bitumen from the tar sands doesn't solve any global problems. Hughes has calculated that one barrel of oil equals eight years of human labour, and he predicts that North Americans will rue the day they squandered their fossil fuels. "It's false logic to think that oil sands will allow business as usual. It's a real mistake." He adds that it's important "we power down and consume less energy. Otherwise Mother Nature will fix the problem, and then it's just a question of how chaotic things will become."

Walter Youngquist, the author of *GeoDestinies,* is a geologist who has worked in seventy countries for companies such as Exxon. Youngquist says that oil has created an incredible hundred-year-long party, but every party, including "the petroleum interlude," must come to an end. Oil production has peaked, and no one has a good plan because none of the alternatives are as portable or versatile as oil. Youngquist describes the tar sands as a last refuge for oil companies and "a valuable long-term resource for Canada," the environmental costs of which are "severe." The Industrial Revolution, he says, has allowed us to exploit our resources at an unprecedented pace, and "its very own success contains within it the seeds of its own destruction." He thinks that Canada should stretch out production in the tar sands for as long as possible, using bitumen efficiently and sparingly. His biggest concern is: "Can we keep the civil part of civilization?"

According to David Finch, an oil-patch historian and author of *Pumped: Everyone's Guide to the Oil Patch,* the average Canadian burns twenty-five barrels of oil a year. The average Albertan burns sixty barrels, due to an above-average use of fossil fuel toys such as ATVs, trucks, and SUVs. The average person in India uses half a barrel annually.

Sixty barrels of synthetic crude add up to a lot of water, earth, pollution, and toxic waste, even using the most conservative numbers. Given the Natural Resouces Canada calculation that it takes an average of three barrels of fresh water to make one barrel of bitumen, I, like other Albertans, am responsible for draining 180 barrels of water (6,303 gallons) from the Athabasca River every year. Using Alberta

Energy's calcuation that it takes two tons of sand to make one barrel of bitumen, my fossil fuel habits now move 120 tons of boreal dirt every year. (That small mountain weighs as much as a blue whale, two thousand people, or twenty-five elephants.) Given the Natural Resources Canada estimation that each barrel of bitumen produces 1.3 barrels of fine tailings toxic waste (most estimates are much higher), I am personally responsible for seventy-eight barrels (or seventy-eight bathtubs) of duck-killing sludge every year. The Uppsala Hydrocarbon Depletion Study Group calculates that it takes fourteen hundred cubic feet of natural gas to make and upgrade one barrel of bitumen. Using that figure, my oil addiction means I crazily consume enough natural gas in a year to heat six homes for a month. The Pembina Institute figures that each barrel of bitumen produces, on average, one ounce of sulfur dioxide (the principal ingredient of acid rain), so I'm making nearly four pounds of acidifying emissions. If, as the Pembina Institute and other agencies calculate, each barrel of synthetic crude produces 187 pounds of carbon dioxide, my fossil-fuel purchases—even before I burn them—have made a five-ton cloud of global warmers. There are five people in my family, so multiply our household damage by five. Nearly twenty million North Americans running on bitumen-based blends have a similar footprint.

Then there are the things you can't measure with numbers. Every time I fill up my tank I'm supporting the First Law of Petropolitics and its corrupt morality; I'm voting with my actions for political integration with the United States and the idea that "the purpose of life is consumption"; I'm driving the so-called nuclear renaissance and subsidizing idiotic notions such as carbon capture and storage; I'm enriching foreign multinationals and state-owned oil companies; I'm displacing farmers in Upgrader Alley and sickening Aboriginal people in Chemical Valley; I'm spending my children's inheritance, because Alberta has saved almost nothing; I'm subsidizing the drug trade in Fort McMurray and bringing cocaine to remote corners of Newfoundland; I'm legitimizing dysfunctional regulators who abuse due process and property rights; I'm killing workers on the Highway to Hell; I'm exterminating woodland caribou, grizzly bears, and boreal songbirds; and I'm accelerating the pace of a global boom in a nation with no vision or goal other than rapid liquidation of a finite resource.

Nobody in the industry, of course, wants to slow down the pace of development. Lynn Zeidler, vice-president of Canadian Natural Resources Ltd. (CNRL), told a public hearing that a slowdown would be both risky and unnecessary. "Actions that erode global competitiveness have and will result in the loss of development opportunities to other countries where more aggressive actions are taken to create environments that facilitate sustainable growth." A representative of Total E&P Canada agrees: "Development in the oil sands is subject to a world class system of governance. We are confident that oil sands development will proceed at a sustainable, appropriate pace, with minimum intervention needed." Although a majority of Albertans favour a slowdown, Alberta's premier, Ed Stelmach, has vowed not to "touch the brakes" on tar sands development.

Herman Kahn thought that hardly anybody would object to the destruction of trees and muskeg in a sparsely populated forest, but he got that part wrong. The United Nations calls the megaproject in the tar sands one of the world's top environmental hot zones, and even U.S. mayors worry about dirty oil and its implications for catastrophic climate change. Innovation Norway, which promotes business opportunities for that country's citizens, bluntly describes tar sands mining as a form of oil extraction "that completely destroys the boreal forest, the bogs, the rivers as well as the natural landscape." Ordinary Norwegians are not impressed. Because the tar sands produce almost the same volume of greenhouse gases as the entire nation of Denmark, a nation of five million people, British activists have called the development "the biggest environmental crime in history." Greenpeace, which has had a field day exposing Alberta's lax environmental standards, set up a satirical website, www.travellingalberta.com, that invites tourists to hang-glide over the mines, sail on the toxic ponds or just sun on "beautiful black sand beaches that stretch for miles."

To counter its new global image as a dirty-oil provider, the Alberta government committed $25 million in 2008 to defending the fiction that rapid tar sands development is "sustainable." As part of that "integrity" campaign, the province sent Alberta's Deputy Minister Ron Stevens to Washington, D.C. in late April with the message that "Alberta represents a stable, secure, and environmentally responsible source of energy supply for the United States." Stevens was greeted with advertisements placed by six major U.S. environmental groups depicting a Canadian maple leaf soaked in tar, as well as headlines about the five hundred ducks dying in Syncrude's tailings pond. Although the ducks made Stevens's trip a public-relations nightmare, he still declared his eye-opening experience in Washington a "mission accomplished."

Well-known continentalist Allan Gotlieb, formerly Canada's ambassador to the U.S., tried to bolster Alberta's sullied image by condemning environmental criticism of the tar sands as "unfair and unwise and potentially damaging to U.S. interests." The Alberta government has also ferried U.S. congressmen to and from the tar sands to prove Energy Minister Mel Knight's claim that "we have not, are not and we will not put our environment at risk for money." The Canadian Association of Petroleum Producers, which rarely appears at public forums, has run full-page advertisements, put corporate leaders in front of the media, and set up its own website to promote "dialogue" about the project's harrowing impact on people, water, and the land (www .canadaoilsands.ca). Marcel Coutu, Syncrude's chairman of the board, contends that industry's "environmental story has been glowing. Where we have done a poor job has been in telling the world about it."

But the environmental debate has barely started. Former Alberta Premier Peter Lougheed predicts the tar sands megaproject will foment an "inevitable" constitutional clash between the federal right to protect the environment and the provincial right to exploit natural resources. Clement Bowman, a former Imperial Oil scientist and oil sands pioneer, echoes that sentiment. Bowman has warned that "the oil sands have almost hit a wall" until the federal government takes seriously the need to clean up the mess in Fort McMurray. Even the *Petroleum Economist* coolly refers to the tar sands as a "foul mess" and an "overrated science experiment." Robert Mansell, a prominent University of Calgary economist, sees other daunting problems ahead. He fears Alberta has put all of its eggs in one tarry, nineteenth-century basket; "substantial climate change, dramatic shifts in future U.S. energy policies or the development of 'game-changing' energy technologies," he writes, could quickly turn the province upside down.

CANADA HASN'T YET had a national debate about the rate and scale of tar sands development and its mammoth implications for sovereignty, water security, the petrodollar, nuclear energy, and climate change. The debate will happen someday, and when it does Canadians will hear essentially two arguments. (As physicist Albert Bartlett noted, "For every Ph.D., there's an equal and opposite Ph.D.")

One group, taking what Bartlett calls the Conservative Path, will hail mostly from the ranks of ecologists, geologists, and retired politicos such as the courageous Peter Lougheed. They'll argue that oil production has peaked and that exploitation of the tar sands without a plan is as about as clever as visiting Mars without a spacesuit. They will warn that unsustainable rates of oil consumption will lead to sustained price hikes and fuel shortages that may, as Walter Youngquist worries, take the "civil" out of civilization. They'll support a persistent, measured decline in oil consumption and recognize population growth as a driver of energy trouble. They will advocate that the federal and Alberta governments cap and limit tar sands production as part of a national energy security strategy. They will likely advocate for a slow but consistent reduction of oil exports too.

On the other side will be the group taking what Bartlett calls the Liberal Path. These will include numerous economists, the National Energy Board, the petroleaders of Canada and the United States, and many tar sands executives. The Liberals don't believe oil will peak for another twenty years. They suspect that all crises are temporary anyway and that tar sands production will soothe some of the world's energy woes. The Liberals believe that the global marketplace, combined with technological innovation, can power any doddery old vehicle. Fossil fuel consumption must rise, because anything else would hurt the economy. The Liberal motto is "dig and drill" anywhere, anytime. The Liberals don't worry about their children, because advanced and as yet undeveloped technologies will save them just in time. They regard population growth as an ally, because "more people equals more brains."

When Albert Bartlett presented these two diverging paths to a U.S. Congressional hearing on energy policy in 2001, he argued that the only rational way for an ordinary person to decide was to compare the outcomes of two possibly wrong choices: "If we choose the Conservative Path that assumes finite resources, and our children later find that resources are really infinite, then no great long-term harm has been done... If we choose the Liberal Path that assumes infinite resources, and our children later find that resources are really finite, then we have left our descendants in deep trouble... There can be no question. The Conservative Path is the prudent path to follow."

To date, Canadian leaders have chosen the Liberal Path, due to willful carelessness or an open disdain for future generations. Many truly believe that bitumen, like hell, has no limits. But the tar sands present Canada with a different opportunity. The unbridled destructiveness there should be a bold invitation for us to live within our means, exercise prudence, and abandon the oil-fuelled mythology of consumption without limits.

TWELVE STEPS TO ENERGY SANITY

. . . .

"We must beat a sustainable retreat."

JAMES LOVELOCK, CLIMATE SCIENTIST, *THE REVENGE OF GAIA*, 2006

MOST PEOPLE KNOW about the famous twelve steps of Alcoholics Anonymous. A society addicted to dirty oil is not much different from a spouse
or parent held hostage by alcohol, even though our governments and
industry promote that addiction as a necessary way to contribute to the
gross domestic product.

It is time for North Americans to seek help. We can ignore the moral
consequences of our addiction no longer. Nor can we afford the dead
end of despair. The following twelve steps would get us headed in the
right direction.

1 ADMIT THE MAGNITUDE AND COMPLEXITY OF THE ENERGY CRISIS.
Cheap oil is a relic of the past. We have undervalued petroleum and
consumed the majority of our fossil fuel inheritance in only sixty
years. Business as usual is an invitation to calamity. The tar sands
are a strategically important Canadian resource that can, if properly
managed, buy the nation a civil transition to a low-oil/low-carbon
economy and be retired by 2030. As Swedish energy expert Kjell
Aleklett puts it: "We have climbed high on the oil ladder and yet we

must descend one way or another. It may be too late for a gentle descent, but there may still be time to build a thick crash mat to cushion the fall."

2 SLOW DOWN TAR SANDS DEVELOPMENT AND CAP PRODUCTION AT TWO MILLION BARRELS A DAY.

The more bitumen Canada produces, the more stuck it will become to industry's decisions and demands. Given the water, energy, and capital intensity of the tar sands, imposing a two-million-barrels-a-day cap would give the governments of Alberta and Canada time to test and regulate cleaner technologies, eliminate toxic tailings ponds, create a long-term plan, and establish real-time reclamation programs. A cap would also put the owners of the resource, Albertans, in control of the pace of development.

3 ESTABLISH A NATIONAL STRATEGY FOR ENERGY SECURITY AND INNOVATION.

Canada is probably the only industrial nation in the world with no clear energy plan and no strategic oil storage for emergencies. A national plan should have many components. It should identify the tar sands as an interim, transitional supply that cannot solve global energy shortages; it should assess the benefits and limitations of renewable energy sources; it should end subsidies for all fossil fuel production; and it should "green" transmission and power plant infrastructure so that businesses and homes can sell green power back to the grid.

4 IMPOSE A CARBON TAX WITH A 100 PER CENT DIVIDEND.

National fossil fuel consumption could be reduced by 50 per cent by 2020 by imposing a series of progressive taxes on carbon, collected at the gas pump. As conceived by NASA scientist Jim Hansen, this carbon tax would be returned in its entirety to the public on a monthly basis. A carbon tax would raise energy prices and the cost of imported food, but ordinary citizens would find ways to reduce emissions in the marketplace by buying energy-saving products that "will spur economic activity and innovation." A well-designed tax system combined with a responsible royalty scheme will ultimately slow consumption, foil fossil fuel lobbyists, and save bitumen for value-added production, such as feed stocks for chemicals and plastics.

5 CHALLENGE THE FIRST LAW OF PETROPOLITICS.
Three simple reforms can subvert the First Law of Petropolitics and the erosion of democratic life fostered by rapid tar sands development.
A *Mandate transparency and freedom of information.*
Governments running on petrodollars must set up more institutions and freedom-of-information tools than non-oil states have. Without these safeguards, citizens will not know where the money goes or how decisions are made. Without greater transparency, bitumen and fossil fuel special interests will resolutely opt for more and more secrecy in political life. Both Alberta and Ottawa need to expand access to information laws and registries.
B *Separate tar sands corporate tax revenues from general revenue to build a national sovereign fund.*
A sovereign fund is a pool of wealth that allows governments to plan for uncertain energy futures. In 2002, the International Monetary Fund recommended that oil-producing states studiously flag non-renewable resource revenues, rigorously watch spending, and dedicate a large portion to a sovereign wealth fund. Norway, Australia, China, Singapore, and the United Arab Emirates all have sovereign funds. To date, Canada has elected to use tar sands cash to reduce taxes. Starting immediately, all federal revenue from the tar sands should be directed to a special savings fund and be managed with the same clarity and integrity as Norway's $400-billion Petroleum/Pension Fund.
C *Reassert accountability in tax regimes.*
Alberta's low royalties have resulted in inflation, reduced competitiveness, and a windfall for the federal government. A higher,

more rational royalty system would encourage more efficient use of capital, slow the pace of development, and foster better project management. Alberta must also reintroduce a real sales tax. Canada must establish a fair income tax that balances the current level of tar sands revenue. The proceeds should be dedicated to renewable energy alternatives and urban infrastructure that will swiftly reduce oil consumption.

6 CHALLENGE CONTINENTAL ENERGY INTEGRATION.

U.S. and Canadian corporate and political leaders want to use rapid tar sands development as the anchor for integrating all energy supplies on the continent. Continental integration assumes that longer and longer global supply lines for hydrocarbons are sustainable and that Canada has cheap energy to spare.

The federal government must rethink these assumptions and adopt "energy autonomy" as a saner option for ordinary Canadians than the shared pain of "energy interdependence." Energy autonomy means extending bitumen pipelines east to Quebec and Atlantic Canada and limiting further exports south. It means transforming Fort McMurray into one of the most energy-smart cities on Earth. It also means recognizing that extending oil, gas, and electricity supply lines to the United States during a period of unprecedented hydrocarbon depletion will multiply continental vulnerabilities and condemn all North Americans to a future of energy insecurity.

7 RELOCALIZE FOOD PRODUCTION.

Cheap oil has created a fantasy food production system that delivers Ugandan peas to Europe and Chinese shrimp to the United States. Canada's agricultural policies, designed during an era of cheap fossil fuels, have largely supported the export of cheap grain and meat. Our nation needs a national food quality and security program that protects fertile farmland; rewards farmers for ecological services such as water conservation; properly labels each product with its origin and its carbon and energy intensity; emphasizes quality, not quantity; favours small operations over big ones; and encourages Canadians to buy locally grown food.

8 ABANDON ECONOMIC DEAD-END ACTIVITIES SUCH AS CARBON CAPTURE AND STORAGE.
CCS is a multibillion-dollar program to accelerate tar sands development that will largely benefit a few oil and gas companies. Instead of spending $16 billion of taxpayers' money on CCS, the Canadian government should make alternative public investments to reduce greenhouse gas emissions, such as funding a new national passenger railway system. Trains, powered by hydro-generated electricity, remain one of the most efficient and pleasant forms of transport for both goods and people.

9 ORIENT ALL RURAL AND URBAN PLANNING TO RENEWABLE ENERGY.
Provinces dependent on hydroelectricity should actively pursue geothermal energy. Provinces dependent on electricity made from hydrocarbons should diversify to wind and solar energy. Instead of building more highways to suburbia, Canada should devote all infrastructure spending to low-carbon alternatives such as walkable communities.

10 PICK THE LOWEST-HANGING FRUIT FIRST.
Canadians—consumers and industry—are so wasteful now that we can make immediate gains in oil conservation in readily observable ways. The Japanese consume only a third of the energy that North Americans do. Europeans enjoy an enviable lifestyle on half of what we consume. A national regulatory program devoted to measuring, auditing, and plugging fugitive emissions from pipelines, wells, and refineries would lower pollution, save billions of dollars' worth of hydrocarbons, and reduce more greenhouse gas emissions in five years than carbon capture and storage could in a decade.

11 DON'T WAIT FOR GOVERNMENT.

Most Canadian governments and political leaders have fallen under the spell of one-dimensional hydrocarbon-thinking and are willing to sacrifice an entire generation of citizens. Power down. Eat local food. Walk more. Travel less. Be a leader in your community and family. Challenge the petrostate.

12 RENEGOTIATE THE NORTH AMERICAN FREE TRADE AGREEMENT.

The aging agreement, now under intense scrutiny by U.S. politicians, currently guarantees the United States unlimited access to Canada's finite oil and natural gas supplies and an unreasonable proportion of oil and natural gas in the event of shortages. As such, it stands as an impediment to "sustainability" for both countries. Albert Bartlett argues that whenever an international agreement "prohibits a nation from following the conservative steps of taking its exploitable natural resources off the international market to save the resources for future domestic use," it is a bad agreement.

Kjell Aleklett at the Uppsala Hydrocarbon Depletion Study Group predicts that global oil supplies will soon become so tight that oil-exporting nations will be forced to "reconsider how much they export and may well save actual reserves for future generations." German energy critic Hermann Scheer argues that "making domestic resources a market priority (and something that leads directly to renewable energy) is sound crisis prevention." The renegotiation of NAFTA is imperative, says Gordon Laxer of the Parkland Institute: "When you cannot safeguard your citizens against freezing in the dark, nor control how much you export, nor set the price at which citizens buy back their own energy from foreign transnational corporations, you know you are not a superpower."

SOURCES AND FURTHER INFORMATION

....

ONE: CANADA'S GREAT RESERVE

Innis, Harold. *The Fur Trade in Canada: An Introduction to Canadian Economic History.* Toronto: University of Toronto Press, 1956.

Mair, Charles. *Through the Mackenzie Basin: An Account of the Signing of Treaty No. 8 and the Scrip Commission, 1899.* Western Canada Reprint Series. Edmonton: University of Alberta Press, 1999.

Murray, Jeffrey. "Hard Bargains—The Making of Treaty 8." *The Archivist* 117 (2000). Available at http://www.collectionscanada.gc.ca/publications/002/015002-2060-e.html.

TWO: IT AIN'T OIL

Alberta Chamber of Resources. *Oil Sands Technology Roadmap: Unlocking the Potential.* Edmonton: Alberta Chamber of Resources, 2006.

Bilkadi, Zayn. "Bitumen—A History." *Aramco World Magazine,* November/December 1984.

Boffetta, Paolo, and Igor Burstyn. "Studies of Carcinogenicity of Bitumen Fumes in Humans." *American Journal of Industrial Medicine* 43 (2003).

Brower, Derek. "In Search of the Key." *Petroleum Economist,* September 2006.

Bruce, Gerald. "Bitumen to Finished Products." Presented at the Canadian Heavy Oil Association Technical Luncheon, Calgary, 9 November 2005.

Ferguson, Barry. *Athabasca Oil Sands: Northern Resource Exploration 1875–1951.* Edmonton: Alberta Culture/Canadian Plains Research Centre, 1985.

Flint, Len. *Bitumen Recovery Technology.* Calgary: Petroleum Technology Alliance Canada, 31 January 2005.

George, Rick. "Creating New Value in the Oil Sands." Presented at the World Heavy Oil Congress, Edmonton, 10 March 2008.

Herron, Hunter E. "Bitumen from Canadian Oil Sands: The World's New Marginal Supply of Oil, Petroleum Equities Inc." Presented to the Rotary Club of Tysons Corner, Vienna, Virginia, January 2006.

Jaremko, Gordon. "Researcher Cracks Secrets of Ugly Bitumen." *Edmonton Journal*, 23 April 2007.

Lyttle, Richard. *Shale Oil and Tar Sands*. New York: Franklin Watts, 1982.

Peachey, Bruce. *Expanding Heavy Oil and Bitumen Resources while Mitigating GHG Emissions and Increasing Sustainability: A Technology Roadmap*. Calgary: Petroleum Technology Alliance Canada, 31 May 2006.

Roy, Jim. *A Review of the Review*. Delta Royalty Consulting Ltd., 19 October 2007.

Van Praet, Nicolas. "Oilsands for Energy 'Bizarre.'" *Financial Post*, 1 May 2008.

THREE: THE VISION OF HERMAN KAHN

Alberta Energy and Utilities Board. *EUB Provincial Surveillance and Compliance Summary 2006*. ST99, 2007.

Alberta Federation of Labour. "Temporary Foreign Workers: Alberta's Disposable Workforce: The Six-month Report of the AFL's Temporary Foreign Worker Advocate." Edmonton: Alberta Federation of Labour, November 2007.

Bodman, Samuel, Secretary of U.S. Department of Energy. Speech at Alberta Government luncheon, Edmonton, 14 July 2006. Available at http://www.doe.gov/news/4059.htm.

Canada. House of Commons. Standing Committee on Natural Resources. *The Oil Sands: Toward Sustainable Development*. Ottawa, 2007.

Canadian Council of Chief Executives. *New Frontiers: Building a 21st Century Canada–United States Partnership in North America*. NASPI report, April 2004. Available at http://www.ceocouncil.ca/en/view/?document_id=365.

———. *Security and Prosperity: Toward a New Canada–United States Partnership in North America*. Profile of the North American Security and Prosperity Initiative (NASPI), January 2003. Available at http://www.ceocouncil.ca/publications/pdf/716af136444 02901250657d4c418a12e/presentations_2003_01_01.pdf.

Carter, Jim. "Oil Sands—The Economic Anchor of the North." Presented at the Meet the North Conference, Edmonton, 9 May 2006.

Cedoz, Frederick. "Everybody Loves Canada... Now." *First Commentary*, September 2003.

———. "Scenarios Suggesting a Stronger U.S.–Canadian Energy Partnership." *First Commentary*, 7 January 2004.

Chastko, Paul. *Developing Alberta's Oil Sands*. Calgary: University of Calgary Press, 2004.

Council on Foreign Relations. *Building a North American Community: Report of an Independent Task Force*. New York: Council on Foreign Relations, 2005. Document available at http://www.cfr.org/content/publications/attachments/NorthAmerica_TF_final.pdf.

Dukert, Joseph. "North American Energy: At Long Last: One Continent." *Occasional Contributions*, October 2005.

Dunbar, R.B. *Existing and Proposed Canadian Commercial Oil Sands Projects*. Calgary: Strategy West Inc., April 2008. Available at http://www.strategywest.com/downloads/StratWest_osProjects.pdf.

Dusseault, Maurice. *Cold Heavy Oil Production with Sand in the Canadian Heavy Oil Industry*. Edmonton: Alberta Energy, March 2002.

E&ETV. "Albert Premier Klein Hits on Tar Sands, Oil and Natural Gas Pipelines, Kyoto Protocol." *On Point* transcript, 29 March 2005. Available at http://www.eenews.net/tv/transcript/55.

Energy Watch Group. *Crude Oil: The Supply Outlook, 2007*. Available at http://www.energywatchgroup.org/fileadmin/global/pdf/EWG_Oil_Exec_Summary_10-2007.pdf.

Flint, Lee. *Workshop Working Report: Oil Sands Experts Group Workshop*. Houston, Texas, 24–25 January 2006.

Fluker, Shaun. "The Jurisdiction of Alberta's Energy and Utilities Board to Consider Broad Socio-Ecological Concerns Associated with Energy Projects." *Alberta Law Review* 42:4 (2005).

Gilbert, Ned. "Interview with Ned Gilbert." *The Tar Paper*, March 2007.

Harper, Stephen. Address by the prime minister at the Canada–U.K. Chamber of Commerce, London, U.K., 14 July 2006.

Hatch, Orrin, U.S. Senator for Utah. "Hatch on Oil: World to Shift Focus to Unconventional Resources." Press release. Woodrow Wilson International Centre for Schools, Canadian Embassy, Washington, D.C., 17 October 2005.

Holroyd, Peggy, Simon Dyer, and Dan Woynillowicz. *Haste Makes Waste: The Need for a New Oil Sands Tenure Regime*. Calgary: Pembina Institute, April 2007.

Humphries, Marc. *North American Oil Sands: History of Development, Prospects for the Future*. Washington, D.C.: Congressional Research Service, 17 January 2008.

Kjell, Aleklett, Bengt Söderbergh, and Fredrik Robelius. "A Crash Program Scenario for the Canadian Oil Sands." *Energy Policy* 35:3 (2007).

Leggett, Jeremy. *The Empty Tank: Oil, Gas, Hot Air and the Coming Global Financial Catastrophe*. New York: Random House, 2005.

McGowan, Gil. Alberta Federation of Labour Presentation to the House of Commons Standing Committee on Citizenship and Immigration, Edmonton, 1 April 2008.

McLean, Archie. "Stelmach Won't 'Brake' Oilsands Growth." *Edmonton Journal*, 5 December 2006.

———. "Work place deaths up 24%." *Edmonton Journal*, 18 April 2008.

Melgar, Lourdes. "Energy Security: A North American Approach." Presented at the North American Forum on Integration, Monterrey, Mexico, 1–2 April 2004.

National Oil Sands Task Force. *The Oil Sands: A New Energy Vision for Canada*, 1995.

Nivola, Pietro. "Energy Independence or Interdependence?: Integrating the North American Energy Market." *The Brookings Review* 20:2 (Spring 2002).

North American Energy Working Group, Security and Prosperity Partnership, and Energy Picture Experts Group. *North America—The Energy Picture II*, January 2006. Available at http://www.pi.energy.gov/documents/NorthAmericaEnergyPictureII.pdf.

Pratt, Larry. *The Tar Sands: Syncrude and the Politics of Oil.* Edmonton: Hurtig Publishers, 1976.

Radke, Doug, Les Lyster, Jillian Flett, and Gary Hayres. *Investing in Our Future: Responding to the Rapid Growth of Oil Sands Development.* Final Report. Edmonton: Alberta Government, 29 December 2006.

Saxton, Jim. *Canadian Oil Sands: A New Force in the World Oil Market.* Washington, D.C.: Joint Economic Committee, U.S. Congress, June 2006.

Smith, Murray. "Alberta's Oil Sands: Turning the Unconventional into the Conventional." Transcription of tape recordings from Interstate Oil and Gas Compact Commission. Austin Annual Meeting. Austin, Texas, 16 October 2006.

————."Canadian Energy and U.S. Energy Security." Presented at the Council of State Governments West Annual Meeting. Breckenridge, Colorado, 10–13 August 2006.

United States. Office of the Press Secretary. *The Security and Prosperity Partnership of North America: Progress,* 31 March 2006.

United States. U.S. Congress. *Canadian Oil Sands: A New Force in the World Oil Market,* 2006.

Wihbey, Paul M. "The Unconventional North American Energy Corridor: A U.S. Senator Makes the Case." *First Insight,* January 2006.

Williams, Cara. "Fuelling the Economy." *Perspectives on Labour and Income,* May 2007.

Woynillowicz, Dan, Chris Severson-Baker, and Marlo Raynolds. *Oil Sands Fever: The Environmental Implications of Canada's Oil Sands Rush.* Drayton Valley: Pembina Institute, November 2005.

Ziff, Paul. "Cross Border Regulatory Collaboration in Its Context: Energy Balances and Energy Policy." In *One Issue, Two Voices: Moving Toward Dialogue; Challenges in Canada–U. S. Energy Trade,* edited by David N. Biette. Washington, D.C.: Woodrow Wilson International Center for Scholars, 2004.

FOUR: HIGHWAY TO HELL

Alberta. Auditor General. *Report of the Auditor General on Alberta Social Housing Corporation-Land Sales System.* Edmonton: Auditor General of Alberta, 20 October 2005.

Alberta Federation of Labour. Presentation to the House of Commons Standing Committee on Immigration, Edmonton, 1 April 2008.

Associated Engineering Alberta Ltd. *Fort McMurray Infrastructure Review: Submission in Support of Intervention of Regional Municipality of Wood Buffalo Application Nos. 1391211 and 1391212.* Fort McMurray: RMWB, July 2006.

Belkin, Douglas. "This Is the Life: Luxurious Digs on Frigid Oil Sands." *Wall Street Journal,* 5 December 2007.

Borsellino, Matt. "Mega Mess." *The Medical Post,* 28 November 2006.

Caldwell, J. "Fort McMurray, Alberta: Oil Sands and Its People Lauded." *I Think Mining,* 25 March 2008. Available at http://ithink.mining.com/2008/03/25/.

Christian, Carol. "CNRL: Review of Deaths Still Underway," *Fort McMurray Today*, 25 March 2008.

Criminal Intelligence Service Alberta. *Annual Report, April 2004–March 2005.* Edmonton: CISA, May 2005.

————. *Report on Organized and Serious Crime.* Edmontòn: CISA, 2007.

The Economist. "Boomtown on a Bender: The Downside of Explosive Growth in Northern Alberta." 28 June 2007.

Edemariam, Aida. "Mud, Sweat and Tears." *Guardian*, 30 October 2007.

Ferguson, Amanda. "Cocaine Easier to Buy Than Pizza." *Edmonton Journal*, 26 August 2007.

Fuller, Alexandra. "Letter from Wyoming: Boomtown Blues." *The New Yorker*, 5 February 2007.

Herridge, Paul. "Alberta Oil Sands Affecting Drug Habits in Newfoundland." *Southern Gazette*, 21 May 2008.

Kohrs, E. "Social Consequences of Boom Growth in Wyoming." Presented at the Rocky Mountain American Association of the Advancement of Science Meeting, Laramie, Wyoming, 24–26 April 1976.

Nichols Applied Management. *Final Report: Mobile Workers in the Wood Buffalo Region.* Fort McMurray: Athabasca Regional Working Group, November 2007.

————. *Sustainable Community Indicators.* Fort McMurray: Athabasca Regional Working Group, January 2006.

Murphy, Verna. "SPCA Over Capacity." *Fort McMurray Today*, 25 March 2008.

O'Leary, Chris. "King Ralph Pays to Twin Highway 63: Alberta's Autobahn." *The Gateway*, 28 February 2006.

Rubinstein, Dan. "Heads in the Sands." *Alberta Views*, March/April 2003.

Sauvé, Michel. "Local Medical Staff Recognizes Community Injury Prevention Initiatives." *Injury Control Alberta* 6:8 (April 2004).

————. *Submission for the Fort McMurray Medical Staff Association.* Alberta Energy and Utilities Board: Canadian Natural Resources Limited Application No 1273113, 20 August 2003.

Seydlitz, Ruth, and S. Laska. *Social and Economic Impacts of Petroleum Boom and Bust Cycles.* Washington, D.C.: U.S. Department of the Interior, June 1994.

FIVE: THE WATER BARONS

Alberta Energy and Utilities Board. *Long Lake Project: Request for Amendment Steam Capacity, Ash Processing Unit & Pad 11, Approval No. 9485*, 22 February 2006.

Alberta Environment and Fisheries and Oceans Canada. *Water Management Framework: Instream Flow Needs and Water Management System for the Lower Athabasca River*, February 2007.

Ayles, G., Monique Dube, and David Rosenberg. *Oil Sands Regional Aquatic Monitoring Program (RAMP) Scientific Peer Review of the Five Year Report (1997–2001).* Fort McMurray: RAMP Steering Committee, 13 February 2004.

Canada. Thirty-ninth Parliament. 2nd Session. Standing Committee on Environment and Sustainable Development. Evidence presented, 18 June 2008.

Canadian Natural Resources Ltd. "Update: Application Filed for Primrose In Situ Oil Sands Project—Primrose East Expansion." March 2006.

Canadian Oil Sands: Prepare for Glory. Scotia Capital: Equity Research Industry Report, October 2007.

Cumming, Joseph, et al. "NAFTA Chapter XI and Canada's Environmental Sovereignty: Investment Flows, Article 1110 and Alberta's Water Act." *University of Toronto Faculty of Law Review,* 22 March 2007.

De Souza, Mike. "Environment Canada Expecting Oilsands Lawsuits." Canwest News Service, 2 March 2008.

Hanel, Joe. "Big Oil Casts Big Shadow over Colorado's Water Future." *Durango Herald,* 6 January 2008.

Mannix, Amy. "Background Report on the Athabasca Basin." Master's thesis, University of Alberta. In press, 2008.

Natural Resources Canada. *Enhancing Resilience in a Changing Climate Program: Water Supply for Canada's Oil Sands,* 2008. Available at http://ess.nrcan.gc.ca/ercc-rrcc/theme1/t7_e.php.

Nelson, Richard. *Strategic Needs for Energy Related Water Use Technology.* Edmonton: Alberta Energy Research Institute, 4 May 2006.

Oil Sands Regional Aquatics Monitoring Program (RAMP): 2000. Volume 1: Chemical and Biological Monitoring. Fort McMurray: RAMP, 2000.

Patton, Wayne, Ian Gates, Tom Harding, Mark Lowey, and Ron Schlenker. *Steam Assisted Gravity Drainage (SAGD): A Unique Alberta Success Story with Implications for Future Investment in Energy Innovation. Paper No. 20 of the Alberta Energy Futures Project.* Calgary: Institute for Sustainable Energy, Environment and Economy, November 2006.

Peachey, Bruce. *Strategic Needs for Energy Related Water Use Technologies: Water and the EnergyINet.* Edmonton: New Paradigm Engineering, February 2005. Available at http://www.aeri.ab.ca/sec/new_res/docs/EnergyINet_and_Water_Feb2005.pdf.

Pratt, Sheila. "Like Marie Lake, Everything Is For Sale in This Oil-rich Province." *Edmonton Journal,* 12 August 2007.

Report of the Rosenberg International Forum on Water Policy to the Ministry of Environment, Province of Alberta. Berkeley: University of California, February 2007.

Robertson, John. "Emerging Technologies and Challenges in Water Use and Re-use in the Heavy Oil Industry." Presented at the Canadian Heavy Oil Association Technical Luncheon, Calgary, 2 October 2007.

Schindler, David, and Vic Adamowicz. *Running Out of Steam? Oil Sands Development and Water Use in the Athabasca River-Watershed: Science and Market-based Solutions.* Toronto: University of Toronto Munk Centre for International Studies, May 2007.

Submission of Giant Grosmont Petroleum Ltd. to the Alberta Energy and Utilities Board Regarding General Bulletin GB 2003-16 Proposed Conservation Policy Affecting Gas Production in Athabasca Wabiska–McMurray Oil Sands Area, 25 June 2003.

Technical Design and Rationale. Fort McMurray: Regional Aquatics Monitoring Program, November 2005.

Bauman, Paul C., and John C. Harshbarger. "Decline in Liver Neoplasma in Wild Brown Bullhead Catfish after Coking Plan Closes and Environmental PAHS Plummet." *Environmental Health Perspectives* 103 (February 1995).

Christian, Carol. "Emergency Debate on Tailings Ponds Denied." *Fort McMurray Today,* 13 May 2008.

Colavecchia, Maria V., Sean M. Backus, Peter Hudson, and Joanne L. Parrott. "Toxicity of Oil Sands to Early Life Stages of Fathead Minnows (*Pimephales promelas*)." *Environmental Toxicology and Chemistry* 23 (July 2004).

Crowe, A.U., A.L. Plant, and A.R. Kermode. "Effects of an Industrial Effluent on Plant Colonization and on the Germination of and Post-germinative Growth of Seeds of Terrestrial and Aquatic Plant Species." *Environmental Pollution* 117 (2002).

Csagoly, Paul, ed. *The Cyanide Spill at Baia Mare, Romania: Before, During and After.* UNEP: The Regional Environmental Centre for Central and Eastern Europe, June 2000.

Dixon, George. *Assessing the Cumulative Impacts of Oil-sands Derived Chemical Mixtures on Aquatic Organisms in Alberta.* Ottawa: Health Canada, 2002. Available at http://www.hc-sc.gc.ca/sr-sr/finance/tsri-irst/proj/cumul-eff/tsri-144-eng.php.

Farrell, A., C.J. Kennedy, and A. Kolok. "Effects of Wastewater from an Oil-sand Refining Operation on Survival, Hematology, Gill Histology, and Swimming of Fathead Minnows." *Canadian Journal of Zoology* 82 (2004).

Fedorak, P.M., D.L. Coy, M.J. Dudas, M.J. Simpson, A.J. Renneberg, and M.D. MacKinnon. "Microbially-mediated Fugitive Gas Production from Oil Sands Tailings and Increased Tailings Densification Rates." *Journal of Environmental Engineering and Science* 2 (2003).

Fekete, Jason. "Duck Deaths Will Hurt Alberta: Harper." *Calgary Herald,* 2 May 2008.

Gentes, Marie-Line, Cheryl Waldner, Zsuzsanna Papp, and Judit E.G. Smits. "Effects of Oil Sands Tailings Compounds and Harsh Weather on Mortality Rates, Growth and Detoxification Efforts in Nestling Tree Swallows (*Tachycineta bicolor*)." *Environmental Pollution* 142 (2006).

Henson, K.L., and E.P. Gallagher. "Glutathione S-Transferase Expression in Pollution-Associated Hepatic Lesions of Brown Bullheads (*Ameiurus nebulosus*) from the Cuyahoga River, Cleveland, Ohio." *Toxicological Sciences* 80 (2004).

Hilton, Carol. "Backlash in Alberta." *The Medical Post,* 6 May 2008.

Jakubick, Alex T., Gord McKenna, and Andy MacG. Robertson. "Stabilization of Soft Tailings Deposits: International Experience." Presented at the Sudbury Mining and the Environment Conference III, Sudbury, 2003.

Klohn, Earle. "Tailings Dams in Canada." *Geotechnical News* Special (1997).

Mackenzie River Basin Board. *State of the Aquatic Ecosystem Report,* 2003.

McKenna, Gord. "Celebrating 25 Years: Syncrude's Geotechnical Review Board." *Geotechnical News,* September 1996.

Meikie, James. "Alarm at Increase in Liver Cancer Deaths." *Guardian,* 17 May 2001.

Morgenstern, Norbert. "Geotechnics and Mine Waste Management–Update." Seminar on Safe Tailings Dam Constructions for the European Commission, Gällivere, Sweden, September 2001.

———. "Oil Sand Geotechnique." *Geotechnical News* Special (1997).

Naumetz, Tim. "Famed Environmentalist Outraged by Criticism of Whistleblowing Alberta Doctor." Canwest News Service, 8 March 2007.

Nix, P.G., and R.W. Martin. "Detoxification and Reclamation of Suncor's Oil Sand Tailings Ponds." *Environmental Toxicology and Water Quality* 7 (2006).

Oiffer, Alexander, James F. Barker, Francoise J.M. Gervais, and David L. Rudolph. "An Investigation of Natural Attenuation of Naphthenic Acids and the Potential for Metals Mobilization at an Oil Sands Mining Facility." Remediation Technologies Symposium, 2006.

Pollet, Ingrid, and Leah I. Bendell-Young. "Amphibians as Indicators of Wetland Quality in Wetlands Formed from Oil Sands Effluent." *Environmental Toxicology and Chemistry* 19 (2000).

Ronconi, Robert, and Colleen Cassady St. Clair. "Efficacy of a Radar-activated On-demand System for Deterring Waterfowl from Oil Sands Tailings Ponds." *Journal of Applied Ecology* 43 (2006).

Stephens, Brett, Chris Langton, and Mike Bowron. *Design of Tailings Dams on Large Pleistocene Channel Deposits: A Case Study—Suncor's South Tailings Pond*, 2005.

Takyi, S., M.H. Rowell, W.B. McGill, and M. Nyborg. *Reclamation and Vegetation of Surface Mined Areas in the Athabasca Tar Sands, Environmental Research Monograph 1977-1*. Edmonton: University of Alberta, 1977.

Taylor-Robinson, S.D., et al. "Increase in Mortality Rates from Intrahepatic Cholangiocarcinoma in England and Wales 1968–1998." *Gut* 48 (2001).

Weber, Bob. "Northern Aboriginals Skeptical of Cancer Study." Canadian Press, 23 July 2006.

Woodford, Peter. "Health Canada Muzzles Oilsands Whistleblower." *National Review of Medicine* 4, 30 March 2007.

SEVEN: THE FICTION OF RECLAMATION

Bacon, Louise. "Creating Wetlands in the Oil Sands, Reclamation Workshop." Presented to the Cumulative Environmental Management Association, Fort McMurray, August 2006.

Brooymans, Hanneke. "Reclaimed Oilsands Site Receives Provincial Blessing." *Edmonton Journal*, 20 March 2008.

Burdeau, Cain. "Did Oil Canals Worsen Katrina's Effects?" *The Huffington Post*, 20 January 2008. Available at http://www.huffingtonpost.com.

Canada. Auditor General. Commissioner of the Environment and Sustainable Development. *Abandoned Mines in the North*. Ottawa: Office of the Auditor General of Canada, October 2002.

Christian, Carol. "Reclamation Certificate Issued." *Fort McMurray Today*, 20 March 2008.

Cowan, W.R., and W.O. Mackasey. *Rehabilitating Abandoned Mines in Canada: A Toolkit of Funding Options.* Ottawa: National Orphaned/Abandoned Mines Initiative, October 2006.

Franklin, Jennifer, S. Renault, C. Croser, J.J. Zwiazek, and M. MacKinnon. "Jack Pine Growth and Elemental Composition Are Affected by Saline Tailings Water." *Journal of Environmental Quality* 31 (2002).

Fraser, Sheila, Auditor General of Canada. "Environmental Auditing." Address to the Cour des comptes of Tunisia, Ottawa, October 13, 2006.

Grant, Jennifer, Simon Dyer, and Dan Woynillowicz. *Fact or Fiction: Oil Sands Reclamation.* Calgary: Pembina Institute, May 2008.

Handel, Steven. *Mountaintop Removal Mining/Valley Fill Environmental Impact Statement Technical Study.* State University of New Jersey, March 2003.

Hanus, Stephen. *Oil Sands Reclamation: Associated Challenges and Mechanism Ensuring Compliance.* Edmonton: University of Alberta, 2004.

Harris, Megan. *Guideline for Wetland Establishment on Reclaimed Oil Sand Leases.* Rev. ed. Fort McMurray: Cumulative Environmental Management Association, December 2007.

Kearl Oil Sands Project: Mine Development, Volume 2: Management Processes. Submitted to Alberta Energy and Utilities Board, July 2005.

Mann, Charles. "How the Energy Business Is Drowning Louisiana." *Fortune,* 16 August 2005.

Mining Association of Canada. *Oil Sands Special Edition, Mining Works for Canada, Summer 2007.* Available at http://www.mining.ca/miningworks/.

Mountaintop Mining/Valley Fill Environmental Impact Statement. EPA/903/R-00/013. Philadelphia: U.S. Environmental Protection Agency, October 2000.

Pauls, Ronald. "Protection with Vexar Cylinders from Damage by Meadow Voles of Tree and Shrub Seedlings in Northeastern Alberta." Proceedings of the Twelfth Vertebrate Pest Conference, Lincoln, Nebraska, 1986.

Polycyclic Aromatic Hydrocarbon (PAH) Monitoring Protocol. Calgary: Canadian Association of Petroleum Producers, December 2004.

Porter, Catherine. "Coal Mining Ravages Appalachia Mountains." *Toronto Star,* 23 February 2008.

Redfield, E., C. Croser, J.J. Zwiazek, M.D. MacKinnon, and C. Qualizza. "Responses of Red-Osier Dogwood to Oil Sands Tailings Treated with Gypsum or Alum." *Journal of Environmental Quality* 32 (2003).

Renault, S., M.D. MacKinnon, and C. Qualizza. "Barley: A Potential Species for Initial Reclamation of Saline Composite Tailings of Oil Sands." *Journal of Environmental Quality* 32 (2003).

Sawatsky L., D.L. Cooper, E. McRoberts, and H. Ferguson. "Strategies for Reclamation of Tailings Impoundments." *Infomine,* November 2006. Available at http://technology.infomine.com/OilSandReclamation/.

Sheppard, Mary Clark, ed. *Oil Sands Scientist: The Letters of Karl A. Clark, 1930–1949.* Edmonton: University of Alberta Press, 1989.

Ward Jr., Ken. "U.S. Judge Cites 'Alarming Cumulative Stream Loss' in Decision." *Charleston Gazette,* 24 March 2007.

Alberta Energy and Utilities Board. Decision 2007-058. North West Upgrading Inc. Application to Construct and Operate an Oil Sands Upgrader in Sturgeon County. Calgary: Alberta Energy and Utilities Board, 7 August 2007.

Alberta Environment. *Cumulative Effects in the Industrial Heartland: A Regional Approach to Managing the Ecosystem.* Edmonton: Alberta Environment, 18 October 2007. Available at http://www.assembly.ab.ca/lao/library/egovdocs/2007/alz/164805.pdf.

———. *The Water Management Framework for the Industrial Heartland and Capital Region.* Edmonton: Alberta Environment, 2007. Available at http://environment. alberta.ca/2276.html.

AMEC Earth and Environmental. *Current and Future Water Use in the North Saskatchewan River Basin.* Edmonton: North Saskatchewan Watershed Alliance, November 2007.

Blake, Donald. *Air Sampling in North-Central Alberta.* Irvine, CA: University of California–Irvine, 26 March 2005.

———. *Air Sampling in North-Central Alberta.* Irvine, CA: University of California–Irvine, 14 April 2007.

Dodds-Roundhill Coal Gasification Project. Public disclosure document. January 2007.

Enbridge Gateway Project. Available at http://www.enbridge.com/gateway.

Griffiths, Mary. *Upgrader Alley: Oil Sands Fever Strikes Edmonton.* Drayton Valley: Pembina Institute, June 2008.

Hawthorne, Michael. "Refinery Pollution May Soar." *Chicago Tribune,* 12 February 2008.

Informetrica Limited. Trans Canada Keystone Pipeline GP Ltd. Application for Construction and Operation of Keystone Pipeline, National Energy Board Hearing Order OH-1-2007.

Jaremko, Deborah. "Operation: Motor City." *Oilsands Review,* March 2008.

Jaremko, Gordon. "Dirty-oil Tag Unfair, Pipeline Chief Says." *Edmonton Journal,* 29 May 2008.

———. "Landowners Take on $2.5-billion Mine." *Edmonton Journal,* 13 January 2008.

Laxer, Gordon, and John Dillon. *Over a Barrel: Exiting from NAFTA's Proportionality Clause.* Edmonton: Parkland Institute, May 2008.

Netzer, David. *Alberta Bitumen Processing Integration Study.* Edmonton: Alberta Energy, March 2006.

Orr, Cameron. "Enbridge Aims for 2014 Start." *Kitimat Northern Sentinel,* 28 May 2008.

Pelley, Janet. "Acid Rain Worries in Western Canada." *Environmental Science and Technology* 40:19 (October 1, 2006).

Shrybman, Steve. Petition to the Governor in Council by the Communication, Energy and Paperworkers Union of Canada (CEP), 13 March 2008.

Stonehouse, Darrell, and Deborah Jaremko. "Bitumen Busters: Upgrading Oil Sands Adds Value, Creates New Markets," in *Alberta Heavy Oil and Oil Sands Guide Book and Directory.* Edmonton: JuneWarren Publishing and the Alberta Government, 2006.

Strathcona County. "Alberta's Industrial Heartland: Oilsands 101 Update." PowerPoint presentation, Sherwood Park, Alberta, 23 June 2007.

Taft, Kevin. "The Heart of a Western Tiger: Fuelling Alberta's Future." Presented at the Rotary Club, Calgary, 11 September 2007.

Wakefield, Benjamin, and Matt Price. *Tar Sands: Feeding U.S. Refinery Expansions with Dirty Fuel.* Washington, D.C.: Environmental Integrity Project, June 2008.

Wu, Nancy, Alberta Employment, Immigration and Industry. "Refining Alberta's Energy Advantage: An Unprecedented Investment Opportunity." PowerPoint presentation. Edmonton, October 2007.

NINE: CARBON: A WEDDING AND A FUNERAL

Anielski, Mark, and Sara Wilson. *The Real Wealth of the Mackenzie Region: Assessing the Natural Capital Values of a Northern Boreal Ecosystem.* Ottawa: Canadian Boreal Initiative, 2007.

Beauregard-Tellier, F. *The Economics of Carbon Capture and Storage* (PRB 05-103B). Ottawa: Parliamentary Information and Research Service, 13 March 2006.

Bramley, Matthew, Derek Neabel, and Dan Woynillowicz. *The Climate Implications of Canada's Oil Sands Development.* Drayton Valley: Pembina Institute, November 2005.

Econenergy Carbon Capture and Storage Task Force. *Canada's Fossil Energy Future: The Way Forward on Carbon Capture and Storage,* 9 January 2008.

Environment Canada. *Environmental Sustainability Indicators, 2007.* Available at http://www.ec.gc.ca/environmentandresources/CESIHL2007/Highlights_e.pdf.

Jaccard, Mark, and Nic Rivers. "Estimating the Effect of the Canadian Government's 2006–2007 Greenhouse Gas Policies." C.D. Howe Institute e-brief, 12 June 2007.

Jaccard, Mark, Nic Rivers, Chris Bataille, Rose Murphy, John Nyboer, and Bryn Sadownik. "Burning Our Money to Warm the Planet: Canada's Ineffective Efforts to Reduce Greenhouse Gas Emissions." *C.D. Howe Institute Commentary* 234 (May 2006).

Kaufman, Stephen. "The Integrated CO_2 Network: A Path Forward for Carbon Capture and Storage." Presented at the Pipeline Conference, Calgary, 28 February 2007.

Peachey, Bruce. "Open Letter to All Stakeholders in Proposed Aquifer Sequestration Projects," March 2008.

Petroleum Technology Alliance Canada. *Review and Update of Methods Used for Air Emissions Leak Detection and Quantification.* Calgary: Petroleum Technology Alliance Canada, 5 February 2007.

———. *Upstream Oil and Gas Fugitive Emissions,* Calgary, 16 March 2006.

Revkin, Andrew. "Is Capturing CO_2 a Pipe Dream?" *New York Times,* 3 February 2008.

Roche, Pat. "Carbonate Klondike: The Next Oilsands?" *New Technology Magazine,* Summer 2006.

Rubin, Jeffrey. "The Efficiency Paradox." *CIBC StrategEcon,* 27 November 2007.

Scheer, Hermann. *Energy Autonomy.* London: Earthscan, 2007.

Science Daily. "Carbon Dioxide Capture and Storage: Grasping at Straws in the Climate Debate." 9 May 2008.

Smil, Vaclav. "Energy at the Crossroads: Scientific Challenges for Energy Research." Presented at the Global Science Forum Conference, Paris, 16–17 May 2006.

———. *Energy Myths and Realities.* In press, 2008.

Statistics Canada. *Human Activity and the Environment: Annual Statistics 2007 and 2008; Section 1: Climate Change in Canada.* Available at http://www.statcan.ca/english/ freepub/ 16-201-XIE/2007000/part1.htm.

U.S. Energy Information Administration. *Country Analysis Briefs: Canada,* April 2006.

Viebahn, Peter, Joachim Nitsch, Manfred Fischedick, Andrea Esken, Dietmar Schuwer, Nikolaus Supersberger, Ulrich Zuberbuhler, and Ottmar Edenhofer. "Comparison of Carbon Capture and Storage with Renewable Energy Technologies Regarding Structural, Economic, and Ecological Aspects in Germany." *International Journal of Greenhouse Gas Control,* January 2007.

TEN: NUKES FOR OIL!

Bersak, A.F., and Andrew C. Kadak. *Integration of Nuclear Energy with Oil Sands Projects for Reduced Greenhouse Gas Emissions and Natural Gas Consumption* (MIT-NES-TR-009). Boston: MIT Centre for Advanced Nuclear Energy Systems, August 2007.

Brower, Derek. "Pushing the Nuclear Button." *Petroleum Economist,* March 2007.

Donnelly, John, and Duane R. Pendergast. "Nuclear Energy in Industry: Application to Oil Production." Presented at the Canadian Nuclear Society's Climate Change and Energy Options Symposium, Ottawa, 19 November 1999.

Dunbar, R.B., and T.W. Sloan. "Does Nuclear Energy Have a Role in the Development of Canada's Oil Sands?" Paper 096 presented at the Canadian International Petroleum Conference, Calgary, 10–12 June 2003.

Ebner, David. "Shell Eyes Nuclear Power in Oil Sands." *Globe and Mail,* 21 May 2007.

Finan, A.E., K. Miu, and A.C. Kadak. *Nuclear Technology and Canadian Oil Sands: Integration of Nuclear Power with In-Situ Oil Extraction* (MIT-NES-TR-005). Boston: MIT Centre for Advanced Nuclear Energy Systems, December 2005.

Fritsche, Uwe. *Comparison of Greenhouse-Gas Emissions and Abatement Cost of Nuclear and Alternative Energy Options from a Life Cycle Perspective.* Darmstadt: Oko-Institut e.v., January 2006.

Hawthorne, Duncan. "What Is the Future of Nuclear Power in Canada?" Presented at the North American Power Markets Conference, Toronto, 29 January 2004.

Henuset, Wayne. "Nuclear's New Frontiers and Canada's Oil Sands." Presented at the Canadian Nuclear Association Annual Seminar, Ottawa, 2007.

Integrated North American Electricity Market—Energy Security: A North American Concern. Toronto: Canadian Electricity Association, March 2007.

Lavallee, Guillaume. "Canada's Oil Sands Going Nuclear." Agent France Press, 26 June 2007.

Lewis, D. Gary. *Oil Sands, Kyoto and the Nuclear Option,* 4 November 2005. Available at http://www.ecolo.org/documents/documents_in_english/Oil-sands-Gary-doc-05.pdf.

Moran, Susan, and Anne Raup. "Uranium Ignited 'Gold Rush' in the West." *New York Times,* 28 March 2007.

Polczer, Shaun. "Nuclear Provider Targets Oilsands." *Calgary Herald,* 26 February 2008.

Rubin, Jeffrey. "Return of the Nukes." *CIBC World Markets: Monthly Indicators,* 17 April 2007.

Rolfe, Brian. "CANDU Plants for Oil Sands Applications." Presented at the IAEA Conference, Oarai, Japan, 16–19 April 2007.

Ross, Elsie. "Nuclear Power Touted as Steam Source for SAGD." *Nickle's Daily Oil Bulletin*, 29 November 2007. Available at http://www.dailyoilbulletin.com.

Scott, Norval. "TransCanada Urges Alberta Linkup to U.S. Power Grid." *Globe and Mail*, 19 December 2007.

Soderbergh, Brent, Fredrik Robelius, and Kjell Aleklett. *A Crash Program Scenario for the Canadian Oil Sands Industry.* Uppsala University: Uppsala Hydrocarbon Depletion Study Group, 6 October 2006.

Solomon, Lawrence. "Burning in the Dark." *Financial Post*, 15 March 2008.

U.S. Department of Energy. *Energy Demands on Water Resources: Report to Congress on Interdependency of Energy and Water,* December 2006.

Winfield, Mark, Alison Jamison, Rich Wong, and Paulina Czajkowski. *Nuclear Power in Canada.* Toronto: Pembina Institute, December 2006.

ELEVEN: THE MONEY

Alberta Royalty Review Panel. *Our Fair Share.* Edmonton: Alberta Government, September 2007.

Anchorage Daily News. "FBI Investigations into Alaska Politics." Available at http://www.adn.com/news/politics/fbi/.

Andrews, Edmund. "Interior Official Assails Agency for Ethics Slide." *New York Times*, 14 September 2006.

———. "Blowing the Whistle On Big Oil." *New York Times*, 3 December 2006.

———. "Inspector Finds Broad Failures in Oil Program." *New York Times*, 26 September 2007.

BBC News (Business). "Exxon Fined $3.4bn for Fraud." 4 May 2001. Available at http://news.bbc.co.uk/1/hi/business/1313246.stm.

Cattaneo, Claudia, and Jon Harding. "Big Investors Concerned about Alberta Royalties Proposal." *National Post*, 4 October 2007.

Chapkro, Evan. "Time to Demand Action on Royalties." *Edmonton Journal*, 13 October 2007.

Devaney, The Honorable Earl E., Inspector General for the U.S. Department of the Interior. Testimony before the U.S. Senate Committee on Energy and Natural Resources, 18 January 2007.

Dunn, Fred. *Annual Report of the Auditor General 2006–2007.* Edmonton: Office of the Auditor General, 2007.

Federal Oil Royalty Underpayment Litigation. Fact Sheet. Washington, D.C.: Project on Government Oversight. 2 January 2002. Available at http://www.pogo.org/p/environment/ea-001028-oil.html.

Finch, David. "The Great Royalty Debate." *Alberta Views*, March 2008.

Francis, Diane. "Beyond Royalty Hysteria." *National Post*, 6 October 2007.

Henton, Darcy. "New Documents Show Royalties Loss Was Billions." *Edmonton Journal*, 5 January 2008.

Hulen, David, and Rich Mauer. "The Alaska Political Corruption Investigation." *Anchorage Daily News*, 7 December 2007. Available at http://community.adn.com/adn/node/112569.

Jacobson, Nathan. "Alberta Oil Sands Report on Threat, Risk and Vulnerability Assessments." Prepared for the Government of Alberta, 16 March 2006. Confidential.

Jaremko, Gordon. "'Give Us Stability': Exxon Chief." *Edmonton Journal*, 8 September 2007.

Lacey, William. "Albertastan? Misguided Intentions and the Fair Share Option." *First Facts*, 19 September 2007. Available at http://www.firstenergy.com.

Loome, Jeremy. "Tories Won't Fund Second AG's Report." *Edmonton Sun*, 30 November 2007

McLean, Archie. "Nuclear Power Lobbyist Ran Tory Election Campaign." *Edmonton Journal*, 22 April 2008.

Markusoff, Jason. "Industry Argument Distorts Picture." *Edmonton Journal*, 27 September 2007.

Morgan, Gwyn. "Populism Tramples Principle in Alberta." *Globe and Mail*, 29 October 2007.

Oil and Gas Royalties: A Comparison of the Share of Revenue Received from Oil and Gas Production by the Federal Government and Other Resource Owners (07-676R). Washington, D.C.: Government Accountability Office, 1 May 2007.

Pitts, Gordon. "Alberta's Tough-love Philanthropist." *Globe and Mail*, 17 December 2007.

Pratt, Sheila. "Klein's 'Don't Worry, Be Happy' Royalty Ideology Doesn't Cut It." *Edmonton Journal*, 23 September 2007.

Rosen, Yereth. "Tales of Oil Industry's Influence in Alaska." *Christian Science Monitor*, 12 December 2007.

Roy, Jim. "A Review of the Review." Delta Royalty Consulting Ltd., 19 October 2007.

———. "Royalty Issues." Delta Royalty Consulting Ltd., 27 June 2008.

Scotton, Geoffrey. "Exxon Warns Against Royalty Change." *Calgary Herald*, 8 September 2007.

Taylor, Amy, and Marlo Raynolds. *Thinking Like an Owner: Overhauling the Royalty and Tax Treatment of Alberta's Oil Sands*. Calgary: Pembina Institute, November 2006.

Thomson, Graham. "Oilpatch Sowing Paranoia and Fear." *Edmonton Journal*, 24 October 2007.

U.S. Department of Energy. *Audit Report: Department of Energy's Receipt of Royalty Oil.* Washington, D.C.: Office of Inspector General, January 2008.

U.S. Department of Justice. *Kerr-McGee to Pay $13 Million to Resolve Oil Royalty Claims More than $275 Million Paid to Date by 10 Companies*, 24 October 2000.

van Meurs, Pedro. "Comments on New Royalties in Alberta." *Calgary Herald*, 27 October 2007.

Vick, Karl. "'I'll Sell My Soul to the Devil.'" *Washington Post*, 12 November 2007.

TWELVE: THE FIRST LAW OF PETROPOLITICS

Canada. Thirty-ninth Parliament, First Session. Standing Committee on International Trade. *Evidence.* 10 May 2007.

Carter, Angela. "Cursed by Oil? Institutions and Environmental Impacts in Alberta's Tar Sands." Presented at the Canadian Political Science Association, Saskatoon, June 2007.

CBC News. "Klein Rejects Environmental Concerns over Oilsands Boom." 4 August 2006.

De Souza, Mike. "Oilsands Penalties Dwarfed by Library Fines." Canwest News Service, 1 July 2008.

Dillon, John, and Ian Thompson. *Pumped Up: How Canada Subsidizes Fossil Fuels at the Expense of Green Alternatives.* Toronto: KAIROS (Canadian Ecumenical Justice Initiatives), 2008.

Friedman, Thomas. "The First Law of Petropolitics." *Foreign Policy,* May/June 2006.

———. "The Power of Green." *New York Times,* 15 April 2007.

Goldberg, Ellis, Erik Wibbels, and Eric Mvukiyehe. "Lessons from Strange Cases: Democracy, Development and the Resource Curse in the U.S. States." *Comparative Political Studies* 41:4–5 (2008).

Johnson, William. *Stephen Harper and the Future of Canada.* Toronto: McClelland & Stewart, 2005.

Klein, Ralph, Brian Tobin, and Gerry Angevine. *A Vision for a Continental Energy Strategy.* Calgary: Fraser Institute, February 2008. Available at http://www.fraserinstitute. org/COMMERCE.WEB/product_files/ContinentalEnergyStrategy2008.pdf.

Kyle, Cassandra. "Klein Visit Spurs Protest."Saskatoon *StarPhoenix,* 24 January 2008.

Martin, Lawrence. "A Prime Minister at the Top of His Imperious Game." *Globe and Mail,* 22 October, 2007.

Nikiforuk, Andrew. "Over a Barrel: Why the Ralph Klein Revolution Isn't Living Up to Its Billing." *National Post Business Magazine,* December 2002.

O'Neil, Peter. "Canada Should Play Big Role in New World Oil Order: IEA." Canwest News Service, 13 July 2008.

Ross, Michael. "Does Oil Hinder Democracy?" *World Politics* 53 (April 2001).

———. "Does Resource Wealth Cause Authoritarian Rule?" Presentation at Yale University, 10 April 2000.

Thomson, Graham. "Turnout Actually Worse than Reported." *Edmonton Journal,* 18 March 2008.

"U.S. Northern Command, Canada Command Establish New Bilateral Civil Assistance Plan." Press release. San Antonio, Texas: U.S. Northern Command, 14 February 2008.

Vulliamy, Ed. "Dark Heart of the American Dream." *The Observer,* 16 June 2002.

Wibbels, Erik, and Ellis Goldberg. *Natural Resources, Development and Democracy: The Quest for Mechanisms,* 2007. Available at http://sitemaker.umich.edu/comparative. speaker.series/files/wibbels_9_07.pdf and http://media.hoover.org/documents/ Wibbels_Stanford_paper2.pdf.

THIRTEEN: EIGHTH WONDER OF THE WORLD

Bartlett, Albert. "Forgotten Fundamentals of the Energy Crisis." *American Journal of Physics* (September 1978).

Bezdek, H. Roger, Robert M. Wendling, and Robert L. Hirsch. *Economic Impacts of U.S. Liquid Fuel Mitigation Options* (DOE/NETL-2006/1237). Pittsburgh: National Energy Technology Laboratory, 8 July 2006.

Cedoz, Frederick. "Oil Sands Rocket Canada to Top." *First Commentary,* 30 May 2003.

Energy Watch Group. *Crude Oil: The Supply Outlook.* EWG-Series No. 3, October 2007. Available at http://www.energywatchgroup.org/Oil-report.32+M5d637b1e38d.0.html.

Farrell, A.E., and A.R. Brandt. "Risks of the Oil Transition." Institute of Physics Publishing, Environmental Research Letter 1 (October–December 2006).

Foreign Affairs and International Trade Canada. *Energy Security: A Canadian Perspective,* 2007. Available at http://www.international.gc.ca/enviro/energy-energie/overview-appercu.aspx?menu_id=20&menu=R.

Gold, Russell. "As Prices Surge, Oil Giants Turn Sludge into Gold." *Wall Street Journal,* 27 March 2006.

Hirsch, Robert L., senior energy program adviser, Science Applications International Corporation. Testimony on peak oil before the House Subcommittee on Energy and Air Quality, Washington, D.C., 7 December 2005.

Hirsch, Robert L., Roger Bezdek, and Robert Wendling. *Peaking of World Oil Production: Impacts, Mitigation and Risk Management.* Report to the U.S. Department of Energy, February 2005.

Hughes, David. "The Energy Sustainability Dilemma: Powering the Future in a Finite World." Presented to the Logan Club Geological Survey of Canada, Ottawa, 25 January 2008. Available at http://aspocanada.ca/david-hughes-presentation.html.

Reguly, Eric. "Oil Sands as an Industry Savior? The Numbers Tell the Real Story." *Globe and Mail,* 12 October 2007.

Robelius, Fredrik. "Giant Oil Fields: The Highway to Oil; Giant Oil Fields and Their Importance for Future Oil Production." Dissertations from the Faculty of Science and Technology, Uppsala University, Sweden, 2007.

U.S. Congress. House Committee on Energy and Commerce. *Understanding the Peak Oil Theory: Hearing Before the Subcommittee on Energy and Air Quality of the Committee on Energy and Commerce* (Serial No. 109-41), 100th Congress, 1st sess., 7 December 2005.

Wihbey, P. Michael. "Global Oil Sands Development and the Rocky Mountain Energy Corridor." Presentation to the Alberta Enterprise Group, 16 January 2008.

FOURTEEN: TAR AGE AHEAD

Alberta Energy. *Talk About Oil Sands,* June 2006. Available at http://www.energy.gov.ab.ca/OilSands/pdfs/FactSheet_OilSands.pdf.

Alberta Government. *Alberta's Oil Sands: Opportunity. Balance,* March 2008.

Aleklett, Kjell. *Peak Oil and the Evolving Strategies of Oil Importing and Exporting Countries.* Discussion Paper No. 2007-17. Paris: Joint Transport Research Centre, December 2007.

Bartlett, Albert. Congressional Testimony on Energy Policy, 3 May 2001. Available at http://www.jclahr.com/bartlett/testimony.html.

Boswell, Randy. "Ottawa Must Reconcile Oilsands Riches and Environmental Challenges: Expert." Canwest News Service, 29 April 2008.

Doucet, Real. "Presentation to Alberta Oil Sands Multi-Stakeholder Committee Consultation Panel." Fort McMurray, 28 March 2007.

Dyer, Simon, Jeremy Moorhouse, Katie Laufenberg, and Rob Powell. *Undermining the Environment: The Oil Sands Report Card.* Edmonton: Pembina Institute, January 2008.

Finch, David. *Pumped: Everyone's Guide to the Oil Patch.* Calgary: Fifth House, 2007.

Hall, Chris, Pradeep Tharakan, John Hallock, Cutler Cleveland, and Michael Jefferson. "Hydrocarbons and the Evolution of Human Culture." *Nature* 426 (November 2003).

Hubbert, M. King. "Exponential Growth as a Transient Phenomenon in Human History." *Focus* 8:1 (1998).

Jaremko, Deborah. "Uncovering the Routes of Civilization in Alberta's Sandbox." *Oilsands Review,* September 2006.

Mansell, L. Robert, and Ron Schlenker. "Energy and the Alberta Economy: Past and Future Impacts and Implications." University of Calgary: Institute for Sustainable Energy, Environment and Economy, 15 December 2006.

McKenzie-Brown, Peter. "Q&A with Marcel Coutu of Syncrude." *Oilsands Review,* 26 June 2008.

Polczer, Shaun. "Oil Sands Will Play Crucial Role in Future: Energy Giants." Canwest News Service, 1 July 2008.

Schramm, Laurier, ed. *Suspensions: Fundamentals and Applications in the Petroleum Industry.* Advances in Chemistry Series 251. Washington, D.C.: American Chemical Society Books, 1996.

Simpson, W.J. "Oil Sands Industry in Fowl Mess," *Petroleum Economist,* 1 July 2008.

Woynillowicz, Dan. *Oil Sands Fever: The Environmental Implications of Canada's Oil Sands Rush.* Drayton Valley: Pembina Institute, November 2005.

Youngquist,Walter. *GeoDestinies: The Inevitable Control of Earth Resources Over Nations and Individuals.* Portland: National Book Company, 1997.

Zeidler, Lynn. "Horizon Construction Management Ltd. Presentation to Alberta Oil Sands Multi-Stakeholder Committee Consultation Panel." Edmonton, 25 September 2006.

TWELVE STEPS TO ENERGY SANITY

Aleklett, Kjell. *Peak Oil and the Evolving Strategies of Oil Importing and Exporting Countries.* Discussion Paper No. 2007-17. Paris: Joint Transport Research Centre, December 2007.

Bartlett, Albert. "A Depletion Protocol for Non-renewable Natural Resources: Australia as an Example." *Natural Resources Research* 15, no 3. September 2006.

Bergevin, Philippe. "Energy Resources: Boon or Curse for the Canadian Economy?" (PRO5 86 E.) Ottawa: Parliamentary Information and Research Service, Library of Parliament. 31 March 2006.

Berry, Wendell. *The Way of Ignorance: And Other Essays.* Berkeley, CA: Shoemaker & Hoard, 2006.

Hansen, Jim. "Carbon Tax and 100% Dividend—No Alligator Shoes." 4 June 2008. Available at http://www.columbia.edu/~jeh1.

Kennedy, Robert, Jr. "The Next President's First Task (A Manifesto)." *Vanity Fair,* May 2008.

Scheer, Hermann. *Energy Autonomy: The Economic, Social and Technological Case for Renewable Energy.* London: Earthscan Publications, 2007.

GENERAL SOURCES

Alberta Energy. *Markets and Pricing for Alberta's Bitumen Production.* Technical Report OS#1, Appendix A. Alberta Royalty Review, 2007.

Isaacs, Eddy, and Duke du Plessis. "Energy Development and Future Outlook." Ottawa: Standing Senate Committee on Energy, the Environment and Natural Resources. 11 December 2006.

Total E&P. Submission to the Oil Sands Multistakeholder Committee (MSC) Consultation Panel: Phase 1, September 2006.

The following individuals were interviewed for the articles that formed the basis of this book: Kjell Aleklett (by e-mail), Matthew Simmons, David Hughes, David Schindler, Randy Mikula, Eric Newell, Michael Ross, Donald Blake, Kevin Timoney, Dr. John O'Connor, Dr. Michel Sauvé, Fred McDonald, David Finch, Alberta Environment Minister Rob Renner, Alberta Energy Minister Mel Knight, Mayor Melissa Blake, Bruce Peachey, Walter Youngquist, Brad Stelfox, Norbert Morgenstern, Jim Roy, Ruth Kleinbub, Grant Henry, Muriel McKay, Fred Dunn, Don Savard, David Rosenberg, Jeff Short, Lee Foote, Anne Brown, Wayne Groot, Dave Sauchyn, and David Keith.

Additional information about the tar sands is available on the following websites:

Alberta Energy
http://www.energy.gov.ab.ca/OilSands/oilsands.asp

Alberta Government: Oil Sands Consultations
http://www.oilsandsconsultations.gov.ab.ca/index.html

Athabasca Regional Issues Working Group
http://www.oilsands.cc/

Canadian Association of Petroleum Producers
http://www.canadasoilsands.ca/en/

Cumulative Environmental Management Association
http://www.cemaonline.ca/component/option,com_frontpage/Itemid,1/

Oil Sands Truth
http://oilsandstruth.org/

Pembina Institute
http://www.oilsandswatch.org

Polaris Institute
http://www.tarsandswatch.org

Regional Aquatics Monitoring Program
http://www.ramp-alberta.org/

APPENDIX:
NORTH AMERICAN
OIL PIPELINES

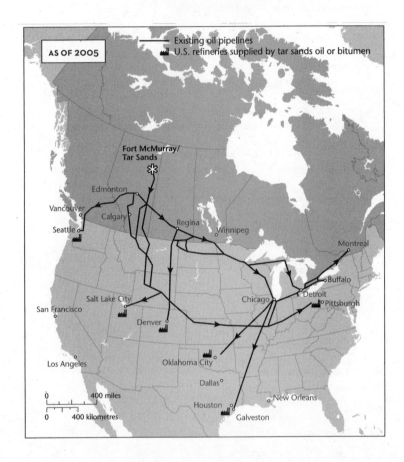

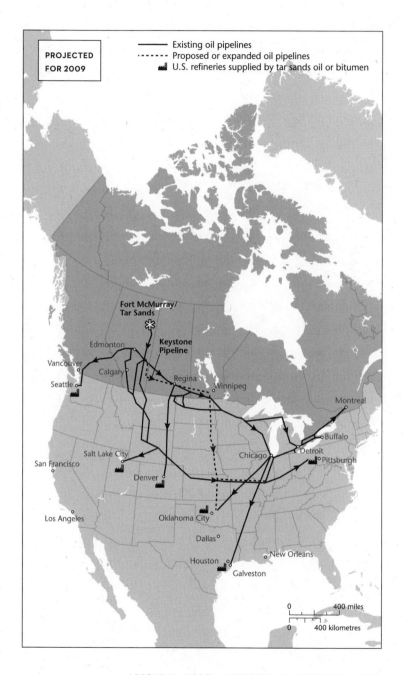

PROJECTED FOR 2009

— Existing oil pipelines
------ Proposed or expanded oil pipelines
🏭 U.S. refineries supplied by tar sands oil or bitumen

Fort McMurray/
Tar Sands

Keystone
Pipeline

Edmonton

Vancouver

Calgary

Seattle

Regina

Winnipeg

Montreal

Salt Lake City

San Francisco

Chicago

Detroit

Buffalo

Pittsburgh

Denver

Los Angeles

Oklahoma City

Dallas

Houston

New Orleans

Galveston

0   400 miles

0   400 kilometres

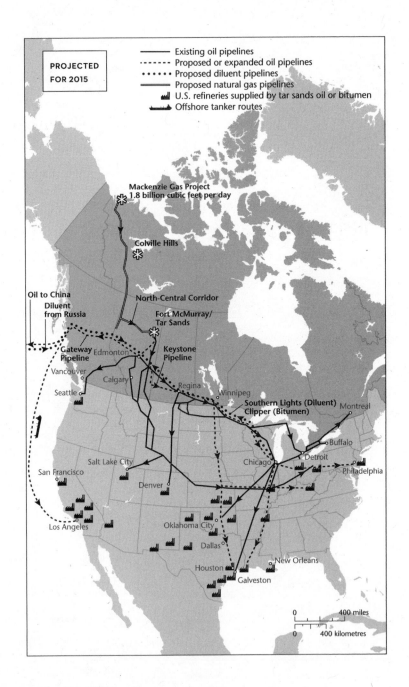

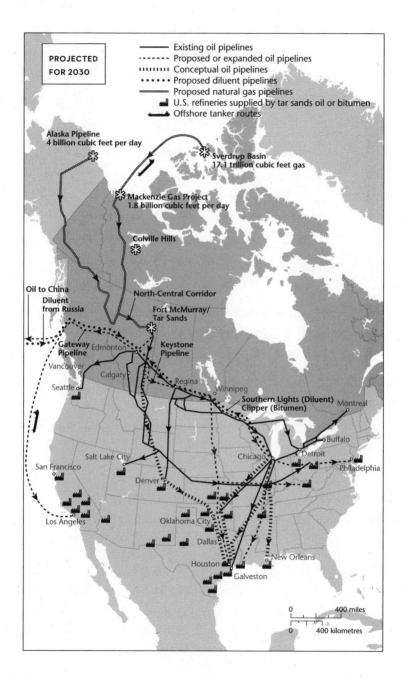

PROJECTED FOR 2030

Existing oil pipelines
Proposed or expanded oil pipelines
Conceptual oil pipelines
Proposed diluent pipelines
Proposed natural gas pipelines
U.S. refineries supplied by tar sands oil or bitumen
Offshore tanker routes

Alaska Pipeline
4 billion cubic feet per day

Sverdrup Basin
17.1 trillion cubic feet gas

Mackenzie Gas Project
1.8 billion cubic feet per day

Colville Hills

Oil to China
Diluent
from Russia

North-Central Corridor

Fort McMurray/
Tar Sands

Gateway
Pipeline

Edmonton

Keystone
Pipeline

Vancouver

Calgary

Seattle

Regina

Winnipeg

Southern Lights (Diluent)
Clipper (Bitumen)

Montreal

Salt Lake City

Chicago

Detroit

Buffalo

Philadelphia

San Francisco

Denver

Oklahoma City

Los Angeles

Dallas

Houston

New Orleans

Galveston

0   400 miles

0   400 kilometres

ACKNOWLEDGEMENTS

EVERY BOOK IS a long intellectual pipeline. First, let me thank a number of editors still committed to critical reporting in a petrostate: Joe Chidley and Scott Steele at *Canadian Business* magazine, Romi Christie and Donna McElligott at the Canadian Broadcasting Corporation, Ted Mumford at *Report on Business* magazine, James Little at *Explore* magazine, and the late Val Ross at the *Globe and Mail*. Their assignments and a presentation to ideaCity created the backbone for this book.

A number of esteemed scientists and individuals generously provided help or corrections for various parts of this dirty saga. They include water ecologist David Schindler; wetlands ecologist Suzanne Bayley; land ecologist Brad Stelfox; physicians Dr. John O'Connor and Dr. Michel Sauvé; energy expert David Hughes; air chemist Donald Blake; geologist Walter Youngquist; oil sands statesman Eric Newell; Pembina researchers Dan Woynillowicz, Peggy Holroyd, and Amy Taylor; historian David Finch; engineer Bruce Peachey; political scientist Michael Ross; statistician Kevin Timoney; and Highway to Hell buddy Oscar Steiner. Many others cannot be named. My deep appreciation to all.

My friends Darrell and Heidi graciously welcomed me into their home in Fort McMurray and shared many insights on life in the boomtown. The irreplaceable Heather Pringle and Geoff Lakeman introduced me to the profane HBO series *Deadwood* and made me a fan of the Al Swearengen school of business. The 2007 Semester in Dialogue class at Simon Fraser University, a burst of youthful energy, gave me the courage to write about what British business journalists now call Canada's Mordor. Petr Cizek ("Tarpit Pete") at oilsandstruth.org graciously shared his excellent pipeline maps.

Under the tightest of deadlines, Barbara Pulling did a masterful edit, while Heather Sangster, Eve Rickert, and Saeko Usukawa weeded out errant commas and grammar. Publisher Rob Sanders kept his promise to deliver a short, snappy book for general readers, and all on schedule.

To my youngest son, Torin Nikiforuk, a most perceptive critic ("You're not working here again, are you?"), I can only say, "Okay. You can have your desk back." I salute my sons Aidan and Keegan for their repeatedly wise advice to a tar-soaked reporter: "Chill out, Dog." To Doreen Docherty, "a wild rose blooming at the edge of the thicket," I offer again my love and gratitude.

Last but not least, historian Harold Innis was right: "The rivers hold sway."

INDEX

steam-assisted gravity drainage (SAGD); wildlife impacts

Environment Canada. *See* Canada, environmental policy

Europe, energy and climate change policy, 123–24, 126, 133–34, 184

extraction processes. *See* open-pit mining; steam-assisted gravity drainage (SAGD)

ExxonMobil, 113, 146, 148

First Nations issues, 6–7, 41, 42, 115, 172–73

fish, impacts on, 59, 61, 62, 63–64, 82, 85, 115

Fort Air Partnership (FAP), 109, 110

Fort Chipewyan, 8, 88–92

Fort McKay, 87–88

Fort McMurray, 8, 10, 27, 39–56, 72, 89, 105, 162

Friedman, Thomas, 153, 166

George, Rick, 13, 31

Gillette Syndrome, 43, 106

greenhouse gases (GHG): Alberta policy, 162; carbon intensity targets, 121–22; carbon tax, 181–82; Energy Independence and Security Act (U.S.), 121; per capita production, 175; from refineries, 115; from tar sands exploitation, 3, 29, 68, 118–28; from uranium preparation, 134. *See also* climate change

Groot, Wayne, 107

Harper, Stephen, 11, 25, 77, 130, 133, 162–63

health issues: cancers and carcinogens, 79, 82, 89–92, 116; in Fort Chipewyan, 8, 88–92; in Fort McKay, 87–88; in Fort McMurray, 49, 51–53; nuclear power and, 135; refineries and, 115–16

Horizon Mine, 61, 80, 98

Hubbert, Marion King, 15, 18, 46, 129, 131, 172

Hughes, Dave, 125, 170, 173–74

Imperial Oil, 29, 46, 53–54, 68, 120 in situ production. *See* steam-assisted gravity drainage (SAGD)

Jean, Brian, 76

Kahn, Herman, 17–21, 28, 171, 176

Kearl project, 29, 59, 61, 97–98, 120

Kennedy, Heather, 159

Khazzoom-Brookes Postulate, 121–22

Klein, Ralph: Heritage Fund, 150–51; as oil "king", 157–58; post-political career, 158–59; promoting tar sands development, 28, 34, 171; royalties and taxes, 140–41, 150, 151

Knight, Mel, 140, 142, 150, 151, 177

Kostuch, Martha, 150

Kott, Pete, 148

Kyoto protocol, 123–24, 162–63

Ladouceur, Raymond, 88–89

Laxer, Gordon, 114, 165–66, 185

litigation, 52, 59, 66

Long, Huey, 156

Lougheed, Peter, 25, 140, 141, 150, 157–58, 177

Louisiana, as petrostate, 99, 156

Mackenzie River Basin, 59–70, 60, 78, 84–85

Mair, Charles, 6–9

Mansell, Robert, 142, 178

Marie Lake, 70

Martin, Laura, 108

McDonald, Fred, 87–88

McEachern, Preston, 64, 66, 85

McKay, Muriel and Steve, 39

Melchin, Greg, 139–40, 150, 159

THE DAVID SUZUKI FOUNDATION

. . . .

THE DAVID SUZUKI FOUNDATION works through science and education to protect the diversity of nature and our quality of life, now and for the future.

With a goal of achieving sustainability within a generation, the Foundation collaborates with scientists, business and industry, academia, government and non-governmental organizations. We seek the best research to provide innovative solutions that will help build a clean, competitive economy that does not threaten the natural services that support all life.

The Foundation is a federally registered independent charity, which is supported with the help of over 50,000 individual donors across Canada and around the world.

We invite you to become a member. For more information on how you can support our work, please contact us:

David Suzuki Foundation
219–2211 West 4th Avenue
Vancouver, British Columbia
Canada v6к 4s2
www.davidsuzuki.org
contact@davidsuzuki.org
Tel: 604-732-4228
Fax: 604-732-0752

Checks can be made payable to The David Suzuki Foundation. All donations are tax-deductible.

Canadian charitable registration: (BN) 12775 6716 RR0001
U.S. charitable registration: #94-3204049

PHOTO CREDIT: DOREEN DOCHERTY

ANDREW NIKIFORUK is a journalist who has written about economics and the environment for the last two decades. His books include *The Fourth Horseman*, which won critical raves in England, Canada, and the United States; *Saboteurs*, which won the Governor General's Award for Nonfiction in 2002; and *Pandemonium*. His work has appeared in numerous publications, including *Saturday Night*, *Maclean's*, *Canadian Business*, *Report on Business*, *Chatelaine*, *Equinox*, *The National Post*, and *The Globe and Mail*, earning him eight National Magazine Awards and an Atkinson Fellowship in Public Policy. He lives with his wife, Doreen, and three sons in Calgary, Alberta.